F*ck Your Feelings

Master Your Mind, Accomplish Any Goal, and Become a More Significant Human

Ryan Munsey

© Strong House Press 2018

ISBN: 978-1090599902

Copyright © 2018 by Ryan Munsey

All rights reserved.
Printed in the United States of America.
First Edition.

For information about permission to reproduce selections from this book, please write to Permissions, Strong House Press, 749 Salem Church Rd, Boyce, VA 22620

For information about special discounts for bulk purchases please email info@ryanmunsey.com or contact Sales, Strong House Press, 749 Salem Church Road, Boyce, Virginia, 22620

Strong House Press. 2018

Table of Contents

PREFACE 1

Why F*ck Your Feelings? 7

   States vs. Traits: Tools in Our Toolbox 9
   Climate vs. Weather 10

CHAPTER 1: F*ck Your Feelings 13

   Can't = Won't 14
   95% of Decisions Are Made
   Based on Feelings 18
   Reflection Questions 28
   Activities 30
   Do It Now 33

**CHAPTER 2: Feelings – Mental
Experiences of Body States** 35

   Feeling Invincible 36
   What Are Feelings and Emotions? 38
   Reflection Questions 50
   Activities 51

**CHAPTER 3: Running the Show –
The Battle Between Our Prefrontal
Cortex and Limbic System** 65

    Marshmallows and the Ulysses Pact … 66
    Reflection Questions … 88
    Activities … 90

CHAPTER 4: Polyvagal Theory and Heart Rate Variability … 107

    Get More Play … 108
    The Polyvagal Theory … 111
    Reflection Questions … 130
    Activities … 146

CHAPTER 5: The Gut – Our Second (or Maybe First) Brain … 159

    The Gut: Our Second (or Maybe First) Brain … 160
    The Chemicals in Our Brain (and Gut) … 167
    Reflection Questions … 197
    Activities … 198

CHAPTER 6: Brain Waves … 211

    That's the Yoga … 212
    Brain Waves … 214
    Reflection Questions … 226
    Activities … 227

CHAPTER 7: Mindset – Upgrade Your Operating System … 241

    Dream (and Fail) Like a Kid … 242
    The Magic Happens Outside of Our Comfort Zone … 247
    Reflection Questions … 255
    Activities … 256

CHAPTER 8: Master Your Actions. Control Your Future. … 267

Faking It 268
Time Management 272
Time Management and Productivity 273
Reflection Questions 302
Activities 320

CHAPTER 9: Habits of High Performers 323

Success Leaves Footprints 324
Reflection Questions 342
Activities 354

APPENDIX 1: My Nine Go-To "Tools" to F*ck My Feelings 357

EIGHT: Float 361

APPENDIX 2: Putting it All Together – The Perfect Day 364

APPENDIX 3: A Few of My Favorite Books and Resources 368

Acknowledgements 370

References 372

About The Author 393

PREFACE

The coldest I have ever been was a beautiful, sunny, 70-degree September day on the beach in New Jersey. While the local beachgoers were sunning themselves and enjoying one of the last picturesque beach days of the year, my teammates and I were peeing ourselves to provide temporary warmth while we shivered and convulsed in a way-too-awkward group hug intended to share body heat.

My teeth-chattering teammates – fellow gym owners, strength coaches, and fitness leaders – and I were several hours into the SEALFIT 20X challenge, a twelve-hour crucible event designed by retired Navy SEALs to help civilians realize how much more (twenty times more) we are truly capable of.

Despite the fact that we dealt with pain, exhaustion, cramps, and yes, self-urination to get warm, the day as a whole remains one of the greatest days of my life and one of my most valuable learning experiences.

It was twelve hours of non-stop physical activity; we were in the ocean, out of the ocean, upside down, carrying logs, and doing all the typical military drills you might imagine but also some you've never heard of (more on those later). We were pushed beyond our physical and mental limits intentionally, in order to teach us that we are capable of so much more than we ever thought.

I have been in a frozen pond in Finland, and I've done snow angels wearing only boxers in the Rocky Mountains, but I can honestly say that nothing has been as uncomfortable, bone-chilling and cold as that SEALFIT day. I distinctly remember that moment – standing on the beach in the sun, with my teeth chattering, shivering, and so tired that I could not move. I could barely pick my head up, and my body was trembling. Every ounce of my body was begging to lay down, seek comfort, and get warm.

But we had more to do. We were about to embark on yet another evolution – what the SEALs called "the segments" of this crucible event – when our instructor saw my hesitation.

He drew a line in the sand with his boot. Literally. He dug his heel into the wet sand and drew a line between us.

He looked me square in the eye and said, "It's your call, Munsey. You don't have to do this. You can quit if you want."

Despite the mind-numbing cold and almost paralyzing fatigue, I stepped over that line without thought or hesitation. I didn't need to think about it. This wasn't a "think about it and get back to me" question. It was a "who are you" question. And despite the discomfort, I knew who I was. I wanted to be that person who saw things through to the end. I wanted to find out how far I could push myself.

That was the moment my life changed. It's a

moment both Coach Mcleod and I will never forget; that was the moment, the decision that forever changed the way I think. That was the moment I became a man who embodies the phrase, "I will do whatever it takes, regardless of how I feel."

"Embody" is the crucial word here. Up until that point in my life, I'd follow through, but only if and when it was convenient. "Whatever it takes" was a value that I applied to my passions – eating healthy and working out – but it wasn't a value that permeated into every aspect of my life.

At some point, if we are to live as the highest expression of ourselves, we all must reach this point. Some reach it at younger ages than others. Many of our crucible moments are decidedly less comfortable than a day at the beach with your friends. Nevertheless, this was *my* moment, and if you've experienced yours, consider yourself lucky, as it is a point of no return. Once we do and become, we cannot un-become.

You may be wondering, if I wasn't there yet, why did I put myself through this SEALFIT 20X challenge?

Good question. First, let me be clear: I have no delusions of being a hero, warrior, or Special Forces operator. I have, however, been fortunate to work with a baker's dozen of these humans on personal and professional levels since that day, and I can say that we now share similar values about mental resiliency.

But the truth is, that wasn't always the case. And the rest of the truth is that my mentors saw things in me I did not see – both strengths and weaknesses. They were the ones who signed up for this 20X event. And a mere four weeks before the big day, when a spot opened up, they invited me to participate. They realized a life experience like this may be just the thing to help me close that frustrating gap between who I was and who I wanted to be.

It was around this time that one of those mentors, Paul Reddick, told me, "Your life is perfectly designed for the results you're currently getting," a quote I'll repeat for you a few times in this book.

I would not be who or where I am today without the amazing people who have served as coaches and mentors throughout my life. Paul, Smitty, Zach, Vince, Uncle Mike, Andres, Brad, and QD on that day, along with too many other amazing people to name, have helped me along the way. I'm grateful for that help, and I hope this book can pay things forward, serving as both a guide and a call to action for you to become the version of yourself that you know you're capable of.

The guidance from my mentors and our SEALFIT experience were a part of the many that laid the foundation for this book. I'm fascinated by and obsessed with our potential as humans – be it special forces operators, Olympic athletes, adventurers, or neuroscientists. Over the last decade, I've studied food science and human nutrition, been a fitness model, owned a performance training facility, helped thousands of people transform their minds and bodies, worked with neuroscientists, interviewed hundreds of experts on behaviors, cognition, evolution, biochemistry, and decision-making – all in a never-ending quest for knowledge – and I have compared the research and theory with the habits of Olympians, Special Forces operators, professional athletes, and entrepreneurs, in search of the footprints that can lead us to success.

What I've learned is in this book. And I can't wait to share it with you.

But before we do that, it's crucial to note that it's not a lack of information that prevents us from reaching our goals. Rather, those who fall short of their goals do so in the implementation or execution stages, despite our stated goals and deepest hopes and desires.

Why? What happens between the time we say we want something and the realization of that effort?

That's exactly what I wanted to answer with this book. Why do some people achieve everything they desire in life, while others seem to stagnate and spend their lives toiling in frustration?

F*ck Your Feelings is an exploration of the neurobiology that makes us who we are. The same biology that is wired for survival and perpetuation of our species can prevent or enable us to achieve greatness. Master it, and you'll write your own future. Let it control us, and we're a raft adrift in the ocean without a rudder, subject to the whims of the tides and breezes.

It's the neuroscience of high performance. Equal parts theory and application, it draws on the research from neurobiology and cognitive behaviorism to explain what feelings are and how they influence our decisions, while simultaneously examining the way high performers are able to successfully navigate their minds, overcome adversity, and achieve their goals.

This book is your guide to self-mastery and optimization.

This is your playbook. It's our user's manual to understand the space between our ears and an arsenal of tools we can implement to master our minds, end self-doubt, build confidence and discipline, become the high performers we want to be, and live as better, more significant, and impactful humans.

The only question left is, what will you accomplish when you master your mind, your feelings, and your actions?

Why F*ck Your Feelings?

A few reasons. First (and probably most poignant for me), it is a phrase I say to myself. It's something I use to alter my state on the days I don't feel like doing the things I know I need to do. It's a phrase that has a deep meaning to me, not only for what it represents, but also for what it has helped me accomplish. I feel like it can do the same for you.

The second reason I chose F*ck Your Feelings is for its ability to jump off the shelf and grab your attention. There are thousands of books you could be reading right now instead of this one. But you're not. You're here; and that means it worked.

I know the message in this book can – and will – change the lives of thousands, hopefully millions. But in order to do that, people have to actually read the book. And before they read it, they have to be aware of it. If I had to risk offending a few people to make sure I had the opportunity to help millions, that was a risk I was willing to take.

The title is memorable and easy for you to share with your friends, family, co-workers, or anyone else you know who will benefit from the information in this book.

Third, F*ck Your Feelings is a relevant title. As we'll find out, 95% of our decisions are made based on feelings. However, the system in our brain that makes these decisions is incapable of processing information outside of the "now." It's the limbic system, and it operates on a mostly subconscious, autopilot level. It's the knee-jerk reaction we have when we see fresh cookies in the window of the bakery.

If we allow this primitive system to run the show, we're significantly less likely to achieve our goals. High performers understand this and have developed their minds, using some of the strategies in this book, to override their primitive biology and accomplish anything.

Finally, F*ck Your Feelings challenges belief systems. It's a jolt – a bit of a shock to your system. It forces you out of that mindless, sleepwalking state and into a state of awareness, straight into the "here and now."

And that's the crux of the book. Being a high performer demands that we move through life with conscious intent and a heightened awareness of ourselves and our brains in real time so that we can make sure each decision we make is aligned with our stated goals.

Are we acting in the *now* in a way that aligns with our long-term goals?

After all, it's our decisions and our choices in life that navigate us to where we are today and where we will go in the future.

States vs. Traits: Tools in Our Toolbox

We have many tools at our disposal to increase emotional resiliency, train our conscious minds to control our primitive and compulsive lizard brains, and ensure decision-making that is congruent with our long-term goals.

In short, these are the tools we can use to F*ck Our Feelings, take back control of our minds, and move our lives in the direction we truly wish to go.

I could have called these "biohacks," but I chose not to. I like to think of them as tools in our toolbox, or weapons in our arsenal.

They're not answers in themselves, but they are modalities we can use to enhance our minds, bodies, and souls. They're not shortcuts, as "biohacks" are often viewed. Rather, these are the lifestyle habits and practices that we can build into and onto our lives. Many of these should become daily, weekly or at least common practices in our lives. The more we practice them, the healthier we become.

They can also be implemented in an acute setting to alter our physiology and provide a positive, short-term, transient state change that is more conducive to our desired goal.

This "now" vs. "lifestyle practice" duality is something that I refer to as "states vs. traits."

Climate vs. Weather

When I first started helping athletes at Clemson and models in New York City back in 2007-2008 with nutrition and personal training, I developed an analogy that I continue to use to this day. I call it the "climate vs. weather" analogy.

If you think about southern California, Florida, or Russia, I'm sure there are certain visuals that come to mind regarding how you might pack if you were to visit one of those areas. Florida and SoCal are known for their sunshine and beautiful climate, but they do have rainy days. Russia generally brings up images of cold, grey skies, bleakness, and snow. But Russia does have warm, sunny days too.

The point with this analogy is that weather is a short-term, daily experience, while climate is an overarching characteristic than outlasts any transient weather pattern. When it comes to our habits, be it working out, eating right, or the emotions, feelings, and thoughts discussed in this book, it's always about the norm, not the exceptions. It's about what you do *most* of the time.

To that end, let's shift from climate and weather to traits and states. Traits are our long-term characteristics – they're our climate or character qualities. They're developed through deliberate, consistent lifestyle choices and actions that eventually become habits, practices, and qualities.

On the other hand, states are short-term, transient experiences. They may last a few seconds or a few days, but they're not permanent. Sticking with our climate vs. weather analogy, states are weather (or moods) as opposed to climate (or demeanor).

As I present tools and activities throughout this book, many can be utilized as both trait- and state-altering mechanisms. There will be some overlap, as much of what we can do "in the moment" should (and will) be a part of our daily routines.

Also keep in mind that most days, our pre-planned (routine) daily version of these habits will suffice to ensure a great day. Other times, however, we may struggle on a given day despite having gone through our normal routine/ritual. This is normal. Be prepared to be flexible and adaptable and be ready to make substitutions or additions depending on the given need and your state (and goals) for that day. These are the days that we "hack" our physiology with the tools at our disposal to F*ck Our Feelings and keep moving forward despite them.

Sometimes, for example, we don't have a workout planned for that day, but simply forcing ourselves to do thirty burpees as quickly as possible can and will change our physiology enough to get us out of whatever funk we were in.

Or perhaps we need to perform our gratitude practice in a unique way to break up the routine and change our physiology. Instead of journaling, maybe you call your grandmother, a parent, or a mentor and thank them for everything they've done for you. Or maybe you thrust yourself in a cold shower, dance to Michael Jackson, or use any of the other physiology shifters we'll outline in this book.

Immerse yourself in these acts. It's about intention, presence, and the ability to shift your focus and get out of your head. After all, the entire purpose of these actions is to positively alter your mental state.

Remember: feelings are mental experiences of our physiological environment. Quite literally, our physiology dictates our feelings. Fortunately for us, we can positively alter our physiology in a matter of seconds with any of these upcoming tactics.

Let's go!

CHAPTER 1:
F*ck Your Feelings

Can't = Won't

It's July 2013, and a mix of adults and pre-teen female softball players are struggling to get off the warehouse floor in near 100-degree heat.

The air is heavy, humid, and filled with grunts, exasperated sighs, and contempt-filled glares. We're at my performance-training facility, House of Strength, and these members are performing ten no-handed burpees.

Before you get any ideas to add these sadistic calisthenics to your fitness program, know that these burpees were *not* supposed to be a part of the workout that day.

These no-handed burpees are something I call "behavior-modification training," and everyone in the facility is floundering around on the ground as penance for one individual's transgression of our most sacred rule. Someone uttered the taboo phrase, "I can't!"

At House of Strength, we had a simple rule: nobody is allowed to say "I can't."

Why? Because I can't = I won't.

This is the lesson I learned from famous copywriter John Carlton. Carlton explains that when we say, "I can't," we're really saying, "I won't do what is necessary to _____."

For example, I can't play the guitar like Bruce Springsteen. But just like The Boss, I'm a human with a brain and ten fingers, and I'm fully capable of developing that ability. But am I willing to put in the hours and years of dedicated practice to master that craft? I'm not.

I can't play like him because *I won't* do what he did to get there.

This lesson resonated with me, so I brought it to the gym. It was permanently affixed to the wall in all capitals: "I CAN'T = I WON'T"

Anytime someone in the gym uttered this taboo phrase, everyone in the facility had to stop what they were doing and do ten no-handed burpees.

Why no-handed burpees? Well, those came from SEALFIT. They were used as a punishment (let's call it behavior modification) for lapses in focus, concentration, or other mental mistakes.

Verbalizing the thought "I can't" is a prime example of such a lapse in mental fortitude and self-belief. "I have an injury or condition that prevents me from safely doing snatches at this time," or, "I have not yet developed the ability to _____" are acceptable alternatives.

"I can't" is a phrase that should be removed from our thought processes and vocabulary altogether. It may seem like mere semantics, but the thought pattern of "I can't" is deadly for those seeking to live as the ultimate version of themselves. It's a sign of a fixed mindset and limiting beliefs. More than nutrition, fitness, strength, or movement coaching, I'm most proud of the role I play in helping others elevate belief systems and possibilities for their life. This was a core value at House of Strength and remains a core value in everything I do.

One of the biggest takeaways after my SEALFIT experience was the knowledge that voluntarily subjecting one's self to adversity could make undesirable, yet required activities less daunting.

In other words, starting our day with a cold shower makes the rest of the day easier because few things will be as uncomfortable as that shower. As we survived that, we feel that we can take on whatever responsibilities or challenges the rest of the day may bring.

Reflecting on the fact that I can do no-handed burpees on concrete fortifies my resolve and helps me believe that I can take on almost challenge thrown my way. A prerequisite to this – and any endeavor in life – is the belief that we *can* do it.

Seeking and overcoming adversity on a regular basis, be it cold showers or no-handed burpees, helps build this confidence and self-belief. Deep down, we must believe in ourselves, or nothing else in this book matters.

It sounds like common sense, but common sense is not always common practice.

High performers realize that it is processes, practices, and habits that achieve results, not people. People perform those actions and get to enjoy the benefits of the traits and qualities derived from those practices, but it is the repeated act of *doing* those practices that produces the high performance and makes one a high performer.

This is why I prefer to coach with a more laid-back approach. No coach does the work for you, no matter how much they yell and scream.

Coaches, mentors, and gurus can only educate and show the way. It's up to us to walk the path and do the work.

Remember the old toy commercial with the disclaimer at the end, "batteries sold separately?" The version of ourselves that we want to be does not rely on others for energy. We've got to be a batteries-included person and bring our own energy to our pursuits.

Now is where we dig in and show you how to start becoming that higher-level version of yourself. We've got to act as if we already are that person. How would they spend their time? How would they handle this problem? How would they organize their day or week? Would they show up and do the work, or would they require the drill sergeant's screams and threats?

The only way to become that version of ourselves is to start acting accordingly. Be that person now. Be that person today and every day. Every damn day. Regardless of how you feel.

As Steven Pressfield says, "If you want to paint, sit your ass in front of the easel."

95% of Decisions Are Made Based on Feelings

You read that correctly: 95% of our decisions are made based on feelings. Not logic. Not rational thought. Feelings. [1]

I remember my reaction when I first encountered that statistic: disbelief. Shock. On one hand, it wasn't a surprise; I knew humans were irrational and emotional, but to see the number – something that came from scientific research – I was startled.

I mean, 95%. That means only 5% of our decisions are made based on facts and logic. I remember my next thought: "If that was true, then our choices are subject to our ever-fleeting feelings and emotions, meaning our lives are a series of emotion-driven events with little forethought, connection, or purpose."

As one who lives by the *Invictus* line, "I am the master of my fate, I am the Captain of my ship," this was a tough pill for me to swallow. Yet it also instantly explained so much. [2]

It explained why I struggled with feelings of frustration and inadequacy from not accomplishing many of the goals on my life's to-do list. There's that word again: *feelings*. I was beginning to see a pattern – feelings leading to choices, choices leading to more feelings, which in turn drove the next choices; it's a circle, and we want to make sure it's an upward cycle, not a downward cycle.

It also explains why we all struggle to see our goals through to completion. We've all felt the supreme motivation and drive to achieve something more with our lives, but for most, that motivation inevitably fades, as does the confidence and drive to achieve that goal.

In fact, only 20% of the population sets goals for themselves.

Of those 20%, only 30% are successful at seeing them through to fruition.

30% of 20% is 6%.

That means only 6% of the population sets and achieves goals. That's startling to me.

I've always been fascinated by why some people achieve so much in their lives while others seem to do just the opposite despite *wanting* to be and do more.

All of this helped explain, in part, why most people find themselves living lives that are not the ones they set out to live five, ten, twenty years ago. For some, like those involved in self-sabotaging behaviors, this is a good detour. But for many, and for the scope of this book, we'll focus on the feelings and subsequent decisions that hinder our ability to realize the highest expression of our existence.

I immediately looked up the researcher behind these findings and dove head-first into the science behind this "feelings drive decisions" phenomenon. It became (and it is) the foundation argument for this book.

Like it or not, our feelings drive our choices and actions. The problem is, emotions and feelings are whimsical and can change more often than the wind or weather. This means that if we're not vigilant over these life-guiding forces, our life paths can be likened to a vessel at sea, completely dependent on the wind and tide – with no way to control our life's own speed or direction.

That's a scary thought for me. And I think it is for you too, otherwise you wouldn't be reading this book. This awareness is critically important for those of us who seek to control our own fate.

Meet Antonio Damasio

Our quest to understand how feelings drive decisions begins with Antonio Damasio, a cognitive neuroscientist at the University of Southern California and author of books such as *Descartes' Error: Emotion, Reason, and the Human Brain* (1994) and *Self Comes to Mind: Constructing the Conscious Brain* (2010). His 2011 TED talk on the Quest to Understand Consciousness was featured on TED's homepage, and he is the head of the Brain and Creativity Institute at USC.

Damasio's research in neuroscience has shown that emotions play a central role in social cognition and decision-making. He is the man behind the "95% of decisions come from feelings" statistic.

One of Damasio's first patients, a formerly successful businessman, father and husband named Elliot, had a tumor removed from the ventromedial prefrontal cortex region of his brain. If you're not a neuroscientist, the ventromedial prefrontal cortex sits at the middle, bottom portion of the brain's frontal lobe (the front of the skull) and is associated with feelings, emotions, decision-making, processing risk and fear, and even morality.

After the removal of his tumor, Elliot began to experience major struggles in his life. Simply put, his life was falling apart because he couldn't make decisions. Even something as simple as choosing a restaurant took all afternoon.

As Damasio explained in his first book *Descartes' Error*, post-surgery Elliot maintained his intelligence, still ranking in the 97th percentile in IQ, yet he was unable to experience emotion. [3]

Damasio said about his interactions with Elliot, "I never saw a tinge of emotion in my many hours of conversation with him: no sadness, no impatience, no frustration."

It wasn't just Elliot having this problem. Damasio discovered that other people with injuries to the brain regions responsible for feelings struggled to make even the most basic decisions.

We might think the inability to experience feelings would make us robotic and facilitate an ideal outcome of making rational, predictable, patterned decisions. However, Damasio observed the exact opposite – a sort of decision-making paralysis.

What is the exact role of feelings in this decision-making process then?

It appeared as though decisions were driven by the feelings generated by somatic markers that originated in the brain, specifically the ventromedial prefrontal cortex and the amygdala. According to Damasio, not only we make decisions based on feelings (both positive and negative), but perhaps more importantly, we often make these decisions non-consciously.

Interestingly, it is that emotional response that links the data to good or bad and drives choices. The lack of emotional impetus therefore keeps the person in an endless cycle of weighing pros and cons, explaining why even the simplest decisions become time-consuming and often insurmountable tasks.

What does this mean for us — those without injuries and full of those feelings and emotions, trying to navigate through life? Good question. As someone who is obsessed with human performance and potential, I'm always looking at new information and trying to figure out how I can apply it to my life (or the lives of those I help) in ways that will lead to increased health, happiness, or performance.

The short answer is this: it means we need to be aware of our feelings and their power to control our choices. It means we need to be aware of the fact that no matter how much we want to achieve something in our lives, it will be our feelings that determine our daily behaviors and ultimately lead to our success or lack thereof, unless we become better managers of our feelings and better decision-makers.

We make thousands of decisions every single day. We all know a single bad decision can have several negative consequences on our life's path. Few of us give enough consideration to the many small, seemingly benign decisions we make on a daily basis because "we don't feel like it?"

I recently read a satirical article exploring what might happen to the food delivery and recipe company Blue Apron if they allowed their feelings to affect the way they showed up for their responsibilities. The articles suggested they watched CNN, got depressed, were frustrated by traffic and decided to forgo their responsibilities in order to stay home, drink wine, and binge-watch Netflix.

As you can imagine, their menu offerings suffered, their creativity went down, and the quality of recipes they provided made their service inferior. In short, they stopped creating the valuable, industry-disrupting service they're known for.

If this played out in real life and continued for multiple cycles of their product, they would lose customers so quickly that their business would be as non-existent as Blockbuster.

This is the exact scenario that plays out in our heads every single day. Every single human being deals with this. It's how our brains are wired.

What fascinates me is the fact that some people seem to overcome this internal struggle and make decisions that bring them closer to reaching their stated goals, while others are driven by feelings and emotions that move them further away from their goals.

For the sake of this book, I'll identify the former group as "high performers" or "successful people," while the latter will be our "normal people."

What makes high performers different? They're not wired differently – at least not from a hardware perspective. It does seem that they're running a different software though – an upgraded operating system if you will.

Before we look at these high performers and how we can implement their proven systems for success into our lives, we must first understand the science of how the brain makes decisions. Once we understand that system, it will be much easier for us to "hack" it to produce the outcomes we seek.

My brain, my personality, and my learning style are such that I prefer (read: need) to see the bigger picture first, then I like to dive into the nitty gritty details. I like to constantly zoom in and out, keeping perspective on how things work as entities but also how they fit into the larger, systemic and holistic picture.

As we explore this central theme, F*ck Your Feelings, we'll start with a high-level overview of the decision-making process, then we'll examine each of the players involved.

I come from a sports background, so I relate to sports analogies. For a team to truly be successful, each player must perform his or her role well as an individual, yet they must do so in harmony with the rest of the team, working together for the common goal. If you prefer a military analogy, simply remove sports and insert military here.

I see the brain and the human body in the same way I see any military or sports team: they are made up of separate components that must properly (and harmoniously) carry out their individual functions in order for the system to function successfully as a single entity.

So what does this team do to make a decision?

How our Brain Makes a Decision

Neuroscientists like Damasio and Ming Hsu remind us that they still don't fully understand how our brains make decisions and there is debate in academia about how it all happens.

For the context of this book and what we need to know about how the decision-making process undermines or supports our goals, we need to understand the limbic system and the prefrontal cortex.

The part of the brain responsible for these compulsive, emotional, feeling-based decisions is the limbic system. It's ancient, fast, and governs most of our subconscious thought systems. It is a primitive, ego-centric, survival-based operating system that is focused only on the "here and now."

The limbic system is not a single area in the brain, rather it is an interlinked system of regions that include the amygdala, hippocampus, thalamus, hypothalamus, basal ganglia, and cingulate gyrus.

As Damasio noted, the amygdala is the emotional center of the brain, and the hippocampus plays an essential role in the formation of new memories about past experiences, linking emotions to memories. [4]

The limbic system plays a crucial role in the formation of memories, integrating emotional states with stored memories of physical sensations. [5]

This is important to note, as it highlights that memories are a tandem of inextricably linked facts and associated emotions. [6, 7]

In contrast, the prefrontal cortex is slower, more effortful, based in logic, future-focused and capable of imaginative and abstract thought, and it is generally regarded as the conscious mind. Where the limbic system is ego-centric, the prefrontal cortex could be considered the "self." It is responsible for about 5% of the decisions we make.

We are going to explore how we can shift ourselves to higher states of consciousness, operating from the prefrontal cortex instead of the lizard-brain limbic system and staying aware of which operating system we're using.

I've been fortunate to interview hundreds of neuroscientists, meditation instructors, brain experts, and consciousness gurus, and the general consensus is that self-awareness is the number one human characteristic we can develop.

If we understand the difference between our conscious and subconscious operating systems, we've laid the foundation to successfully begin the process of shifting to increased awareness and high states of consciousness – not to mention, making ourselves more impervious to those pesky feelings that trip us up on our path to our authentic life.

Decision-Making Is More Complex Than We Realize

I wish I could present you a diagram that showed each part of the brain and the order in which each area is accessed to produce a decision step by step.

Since decisions are all unique, and most contain a series of other decisions, the decision-making process is less of a one-direction flow chart and more of a cyclical feedback loop with as many entrance and exit points as an LA freeway.

Even the simplest decisions require many complex considerations, often involving different regions of the brain.

For example, consider the simple decision of where to eat for lunch that Damasio's patient Elliot struggled with. These are some of the factors involved in such a seemingly simple decision:

- How hungry is he?
- What kind of food does he want?
- Who is he eating with?
- What might they want?
- Do they have dietary restrictions?
- Are we confined to establishments within walking distance or can we drive?
- How much time do we have for this meal?

There is more planning, processing, and comparing involved in any decision than we are consciously aware of at first. I'm sure you can add a few more considerations to this list of questions.

As you can see, there are considerations of time, empathy, morality, and social constructs, all of which are regulated in distinct brain regions. Without the ability to "feel" one way or another about each of these considerations, Elliot was unable to get out of an endless loop of weighing the pros and cons of each.

This communication between brain regions is conducted by neurons, while neurotransmitters function as the chemical messengers that transmit information between neurons.

Before we take an in-depth look at neurotransmitters, we need to explore our feelings. So grab a box of tissues and get comfortable on the stereotypical psychotherapist's couch.

Reflection Questions

Each chapter of this book will include sections for reflection questions and tools to help you implement the information presented in that chapter.

The reflection questions are designed to give a jumping-off point for introspection. You'll hear me say this several times, so one more won't hurt: information without implementation is wasted knowledge – perhaps worse.

The sections with reflection questions are your call to action. They work best if you pause and give them some thoughtful consideration.

You can write your responses in a journal or you can simply mull these over in your head, however you do your best thinking.

When I read books, I take notes in Evernote. If I'm listening to audio while driving or walking, I voice-dictate the note. If I'm reading a physical copy of a book, I type the notes. Either way, Evernote syncs across my devices, and my notes are always handy.

1. When have you let your feelings dictate your choices? Do you notice recurring themes or patterns?
2. How often do you change plans to work out, eat healthy, pursue your work or passion project because "you don't feel like it?"

3. Are you on the life path you truly want for yourself?
4. If not, what does that path look like? What is one immediate action step you can take to get on that path?
5. If you are on that path, what does the next step or chapter look like? What can you do to continue to grow, evolve, and progress?

Activities

What are the strategies to shut off the voice in our head that tries to take over our decision-making process? The key is awareness – being aware of what is happening and realizing that, "Hey, this is not me; this is not something that is in alignment with my values and my goals." If you can have that awareness, you can separate that from yourself. Usually, that is coming from the limbic system, and it is a primal or primitive urge to satisfy something – some physiological state.

Falling back on neuroscience again, we know that physiological states create mental states, and those are our emotions and our feelings. So if you are feeling or thinking a certain way, but you know you may not have the energy to go for a run or do some push-ups or squats to change your physiology, just changing your posture (moving from a hunched-over posture of depression to a posture that is erect, open and safe) can go a long way. Coupling that with awareness, those are some of the things that are going to work in the long term. But in the short term, you are going to need some "hacks."

Distract yourself: watch your favorite movie, listen to your favorite music or talk to somebody. One of my favorite things is to do something for someone else, whether it is volunteering locally or just doing something in a matter of five minutes. Pull out your phone, text five different people and try to make their day. That will change your physiological and mental state.

A lot of it depends on what those thoughts are in your head, but again, the key is being aware of that. If you can be aware and identify that it is something that is not in alignment with who you are and where you want to be, it lets you off the hook, like it is not you.

Realize it is a primitive calling, and just because it is there, it does not mean that it has to be answered. You can use the higher portion of your brain to decipher where it is coming from and why it is there, and if that is too much effort, then just accept that it is there and say, "No, I am not going to listen to it."

#MTC – Move the Mother F*cking Chains

I'm a big list person. I have lists on my phone, computer, in my office, and in the kitchen. My mind is always churning on something, and I have to write things down if they're ever going to have a chance to survive.

One of those lists is a yellow legal pad that serves as my to-do list for each day. The top right quadrant of each page has #MTC, which stands for "Move the Chains." Under that, I write the single most important activity for that day. This is the one thing that needs to get accomplished on that day, no matter what else happens.

If at all possible, I attack that item first thing in the morning. Sometimes it's an appointment, a podcast, or a phone call that has to wait – in which case, the timing is out of my control.

I know that accomplishing that one item will move my mission forward, and I make sure that one thing gets done with my full presence and intention.

Why do I call this moving the chains? Credit for that goes to my mentor Paul Reddick. Paul used the football analogy to instill in me (and hundreds of others) the simple, one-step-at-a-time process to ensuring continued progress.

In football, you get four plays to gain ten yards, or a first down. Every first down resets the four plays to get ten more yards. Most teams will punt on fourth down, so we divided ten by three to get 3.3, rounded up to 3.4 yards. If a team could gain a mere 3.4 yards on every single play, they would never punt. They would score touchdowns on every possession. And putting points on the board is the name of the game. The team with the most points on the board has yet to lose a game.

The lesson here is to forget about Hail Marys, flashy and fancy highlight reel plays. Run it up the gut, gain 3.4 yards per play every day and keep moving the chains on your mission and your life. It is amazing how powerful this practice is. It reminds me of *The Compound Effect* by Darren Hardy.

It's also amazing how hard I am to tackle at the three-yard mark. In other words, that single action builds momentum, and I usually end up achieving well beyond the bare minimum of 3.4 yards per play (or day). This, more than anything else I do, has contributed to my progress and work output over the last five years.

The important part of this activity is to make moving the chains a habit and a trait that is state-independent. In other words, something you do no matter how you feel. F*ck Your Feelings. Do it anyway. That's the real secret of high performers.

Do It Now

Another trick to develop a bias toward action is the "do it now" rubber band trick. Our thought process should always be "take immediate action and do it now." To help make this a habit, wear a rubber band on your wrist, and anytime you feel yourself avoiding a phone call, not answering an email or taking out the trash, snap that rubber band and say, "Do it now!"

If you're thinking these tasks are too small to make a difference, you should know that it's all about habits, practices, and wiring. This is a powerful tool to help us become a person of action. Start small, develop habits that become second nature, and you'll be surprised at how much progress you've made in two, six, or twelve months.

An Important Note on Bad Days

We all have one of those from time to time. That's why I love the #MTC on my to-do list. On a bad day, I'll do that *one* thing, then shut it down and spend time in recovery/recharge mode, focusing on self-care.

I've achieved my proverbial 3.4 yards for the day, I have kept my mission on pace, and I can then focus on making sure the next day is not another bad day.

Give yourself permission to just be – be still, sad, tired, whatever it is.

Think big picture. Remember: it's about the climate, not the weather.

Get your 3.4 yards in for the day and then take care of yourself.

CHAPTER 2:
Feelings – Mental Experiences of Body States

Feeling Invincible

Picture this. You're at a Tony Robbins seminar. The master of motivation is doing his thing, and the crowd is transformed. The place is electric, and you're feeling empowered, motivated – even invincible. In this moment, there is nothing you can't do.

It doesn't have to be a Tony Robbins event. Maybe it's a conference that celebrates your passion – Paleo f(x), Podcast Movement, the CrossFit Games – the specific event matters less than the immersion in the tribe and the moment.

You've likely experienced such a moment surrounded by your favorite athletes, authors, speakers, or industry leaders along with hundreds of passionate, supportive, like-minded individuals. If you've been there, you know that feeling of invincibility we get. That feeling that nothing is out of your reach. That feeling that you can take on any challenge and crush it.

Why then are we back to our "normal" selves by Wednesday or Thursday, when we get home? Why don't those feelings and emotions last? Our values haven't changed. Our goals haven't changed since we left the event. We are no less committed to the pursuit a mere seventy-two hours later. Why has Superman become Clark Kent again?

This chapter will explore the difference between feelings and emotions, their evolutionary roles, and how we can control them so they don't control us.

But first, a caution against the peak states we've all experienced at these events. By definition, peaks don't last. By definition, a peak is followed by a drop-off.

Rather than seek peak states or peak performance, we should seek sustainable levels of high performance that do not result in crashes or drop-offs.

Successful people know that the journey is a long one and relies on consistency and longevity that outlast any single feeling, emotion, or state. Peak states, no matter how great, are just that: short-term, transient states.

Being a high performer, on the other hand, is a trait that we can tap into daily and for the long haul. We don't need to be Superman to achieve the things we want in this life. We simply need to show up, do the work, and be consistent.

Sure, there are things we can do to elevate our states and improve productivity (we'll explore many of them), but our pursuit should be to develop ourselves into a higher-level human that we can maintain on a daily basis, not one that requires the perfect setting to reach a certain temporary altitude only to fall flat later and constantly need to chase the next high.

Successful people realize that processes get results, not people. They show up, consistently, and do the work required, no matter how they feel. They're masters of saying "f*ck you" to their feelings.

What Are Feelings and Emotions?

Before we can fully grasp the way in which our feelings shape our decisions, and ultimately the trajectory of our lives, we must first understand what they are and why we're biologically wired to act based on our feelings.

According to Antonio Damasio, feelings are "mental experiences of the body states." Knowing that our physiology (or body state) drives feelings is very powerful information. That alone tells us that changing our physiology can alter our feelings.

From his neurobiological perspective, Damasio says feelings arise from our emotions. Specifically, feelings "arise as the brain interprets emotions, themselves physical states arising from the body's responses to external stimuli."

Did you catch that?

- External stimuli create physiological responses.
- These physical states, emotions, are interpreted by the brain, giving rise to feelings.
- Feelings, in turn, drive decisions.

From this point forward, the terms "feelings" and "emotions" will not be used interchangeably. They are separate, yet related neurobiological events.

What are emotions and how are they different

from feelings? In the quote above, Damasio described emotions as physical states based on external stimuli.

Emotions are the physiological states that we experience as feelings in our heads.

As I dug deeper, I encountered research from Dr. Adam Anderson, a Behavioral Neuroscientist at Cornell University. The first study I read from Dr. Anderson was a 2005 paper stating that emotions serve to shape our focus. Like Damasio, he notes that we collect input from external stimuli and send that data up the chain of command, if you will, as emotions.

He wrote, "emotions thus bias attention to focus on events associated with subjective and physiologic arousal, shaping the ultimate contents of awareness." [8]

Looking for clarification, I reached out to Dr. Anderson and I am indebted to him, as he was a tremendous resource for me as I worked to connect the dots on feelings, emotions, and our decision-making process.

I asked him to *define emotions from a neurobiological perspective*. His reply highlights the different beliefs in academia, as both sides have data to support their stance – further evidence that this domain of scientific exploration is in its infancy (adolescence at best):

"There is a lot of academic argument over what is an emotion vs. affect vs. feeling. I think it's mostly academic in the worst sense of that term, meaning it might not reflect reality. Typically, emotions are specific categories that we're all familiar with, having distinct expressions like fear, anger, sadness, joy etc. We call these the 'basic' emotions: a small set that are recombined for more complex states. [There exists] a lot of debate of whether these have some evolutionary genetic origin or are socially constructed (i.e., learned in our culture). Affect refers to the underlying valence (positive/negative, good vs. bad) and arousal components of experience. Some researchers believe that the brain only generates affect, and emotions are just words we use for affect in different contexts. In this way, there would be no 'fear' region of the human brain. Obviously, there is a lot of disagreement over this.

My work fits within a model of emotions as specific adaptations all meant to help us, like different tools developed for special tasks. Even the emotions that feel 'bad' are meant to be good for us. That is, they are good for us (survival advantage) when experienced in moderation and when appropriate for the situation. When those later conditions are not met, emotions become maladaptive. You don't want to remove someone's capacity for fear or anxiety; you want to harness it towards optimal function. At the extreme, loss of the functional aspects of emotions results in disorders (depression, generalized anxiety). But evolution did not discover sadness and fear to decrease our ability to pass on our genes." [9]

There are a lot of takeaways from his words:

- Emotions can have evolutionary genetic origins and/or be socially constructed.
- Some researchers believe that the brain does not have specific regions for "fear" or other emotions, but only the underlying valence

(good/bad); this is hotly debated.
- Emotions should be seen as tools developed for specific tasks. Even emotions that are "bad" serve a purpose, especially fear in a survival situation.

Like any tool used inappropriately, emotions that are not regulated properly (dysfunctional) can have unintended, negative outcomes.

I think our emotional states are a combination of societal constructs and evolution. It's difficult to completely dismiss the belief systems that frame our subconscious thought, yet I also agree with Dr. Anderson's statement that "evolution did not discover sadness and fear to decrease our ability to pass on our genes."

Again, I asked Dr. Anderson for clarification: *"How does one distinguish between emotions and feelings?"*

"Feelings are the embodied experiential aspect of emotions, resulting in sensory experiences. Unlike thoughts, emotions come with feelings, making them a special category of mental experience. Much of the more biologically inclined of us follow the thinking of William James (originator of the somatic marker hypothesis), thinking these feelings come from interceptive activity, originating from how bodily states are relayed to the brain. We have been working a lot on this issue, demonstrating how much our brain knows about heart dynamics, for instance. In this way, the dynamics of the heart, breathing, blood flow become sensory signs from the inside of the body sent to the brain, similar to how the brain reads visual information from the eyes. Just turns out we have much better visual acuity! So our internal feels are much more diffuse, making them hard to know well, despite their power on our decision-making."

The somatic marker hypothesis, mentioned here by Anderson and first attributed to William James, proposes that certain feelings are linked to specific emotions, and that these underlying stimuli drive our decisions. [10]

An example is the feeling of a rapid heartbeat being linked to emotions like anxiety, fear, panic, or stress. The theory posits that subsequent decisions are affected by the acute change in emotion through a process that involves an interplay between neural systems that elicit emotional/bodily states and neural systems that map these states.

Somatic markers – things like elevated heartbeats or nausea – are thought to be processed in the ventromedial prefrontal cortex and the amygdala through the processes that we have outlined up to this point.

One of the first to identify this neurovisceral connection in the 1880s was William James, a Harvard trained Physician and Psychologist fascinated by belief systems and what he called "biological fitness."

Influenced by his godfather, Ralph Waldo Emerson, James also taught at Harvard, and some of his illustrious students included W.E.B, Du Bois, and Theodore Roosevelt. In his original somatic marker theory, James argued that "physiological arousal instigates the experience of emotion." [11]

Meanwhile, in 1885, a Danish Physician named

Carl Lange independently developed a similar theory, stipulating that "all emotions are developed from, and can be reduced to, physiological reactions to stimuli." [12] Lange also published work on depression that formed the foundations of what is known today as major depressive disorder.

The ideas of these pioneers have collectively been known as the James-Lange theory, which is widely regarded as the origin of the somatic-marker theory, "based on the central concept that stimuli that induce emotions such as fear, anger or love initially induce changes in visceral function through autonomic nervous system output, and that the afferent feedback of these peripheral changes to the brain is essential in the generation of specific emotional feeling." [13]

In other words, the brain is influenced by the body and vice versa.

This is great news for anyone battling down days, depression, blues or emotional states they wish to change. It really is as simple as changing our physiology.

Since the original thoughts were published, the James-Lange theory has been studied, criticized, modified, adapted, and advanced; however, it is undeniable that this was the foundation of an ever-evolving field of study.

Before we continue, I don't want to overlook Anderson's statement that our heart's dynamics impact the brain and its function. The James-Lange theory posits a neurovisceral link, and we'll dive deeper into this with Dr. Stephen Porges and his polyvagal theory in Chapter 4, but here's a hint: increased vagal tone (synonymous with heart rate variability) increases emotional resiliency.

Emotions Drive Focus: Prioritize Information Processing

With the sheer volume of information coming into our brain at any one time, and throughout each day, it's impossible to give each data point our full attention. This was true thousands of years ago, and it's even more relevant in today's world.

That's where our emotions come in handy. They're our brain's way of signaling "red flag" when we encounter something in the environment that it wants us to focus on.

It's up to us to give that emotion some attention (awareness) and to try to figure out what has stimulated this physiological response.

In short, emotions are designed to collect data from our experience of the outside world and direct our actions. Contrast this with feelings, which are the mental states associated with these emotional responses.

For such a complex brain, this all sounds kind of simple and primitive, doesn't it? That's because it is. We haven't always had this massive brain, and many of our default settings are rooted in the primitive, survival-focused sections of our brains.

One of those primitive processes involves the limbic system. Also called the paleomammalian cortex, it is among the oldest systems in our human brain, making it no coincidence that people with heightened emotional maturity and emotional intelligence are labeled "evolved."

The limbic system's amygdala is responsible for linking incoming data-based emotions (framed by evolution and societal constructs) to the mental states we call feelings.

A meta-analysis of papers published by Anderson and some of his behavioral neuroscientist colleagues explains the amygdala's role like this:

"The amygdala may thus represent embodied attention – the crucial link between central (mental) and peripheral (bodily) resources. Attention and awareness should be coupled with information of sufficient importance to result in autonomic activation and associated mobilization of peripheral metabolic resources, such as through altered respiration and heart rate. This redirecting of bodily physiologic resources may serve as the common primitive foundations of both attention and emotional experience."

It is important to note that when activated, the limbic system channels "metabolic resources" *away* from the brain and towards the extremities for action. This crucial acknowledgement will prove relevant when we discuss awareness and the importance of training our brain to shift from subconscious thought to conscious thought. They continued:

"This common origin supports why emotional states are strongly correlated with how we think about and perceive the environment." [14, 15, 16, 17]

What we need to take away from this now is that the limbic system's role is to link mental states (feelings) to the physiological state (emotion) created by the environment to help direct our focus and determine our next action (choice).

Why Are We Wired This Way?

For their part, emotions were never designed to be precise; they were designed to be fast, for survival. This may be why we're encouraged to "think twice," "sleep on it," or "avoid knee-jerk reactions" when making major decisions. We don't want to make potentially life-altering decisions using primitive pathways designed for survival, not higher-level cognition.

The limbic system predates our language system, and according to Paul D. MacLean's triune brain theory, it is involved with the fight or flight response of the sympathetic nervous system. [18]

MacLean, who passed away in 2007 at the age of 94, was a pioneer in neuroscience, positing that the brain is actually three brains in one: the reptilian (lizard) brain, the limbic system, and the neocortex.

In MacLean's theory, the neocortex – with the prefix "neo" for "new," signifying that it was the most recently evolved area of the brain – was where superior mammalian processing and higher-level consciousness, intelligence and abstract thought took place. You'll see this again soon.

MacLean noted that these three "brains" were interconnected, yet retained "their peculiar types of intelligence, subjectivity, sense of time and space, memory, mobility and other less specific functions."

In other words, like our team analogy in the beginning of this book, they worked independently but synergistically when functioning properly.

We could also use the analogy of parallel universes. Like parallel universes inside our mind, we have multiple competing systems that, at all times, present us dichotomous and paradoxical choices. Choices that, when approached with increased awareness, can help us align our actions with our long-term goals and accomplish anything, no matter how we feel in any given moment.

MacLean wasn't the first to discuss the primitive, evolutionary nature of the limbic system and our emotions. Both Charles Darwin and Robert Dawkins wrote about the evolutionary purpose of emotions two centuries ago. [19, 20]

Darwin speculated that the emotions must be the key to the survival of the fittest, while Dawkins noted that if "emotions are widespread across both human and animal kingdoms, they have been proved, evolutionarily, as crucial to the process of survival, and are inextricably linked to the origins of the species."

We must remember that emotions are quick – extremely quick. They're faster than thought and provide us with the ability to make rapid, sound decisions in complex and uncertain situations, giving us an evolutionary advantage.

Let's say you and a loved one (a child, perhaps) are walking through the woods, and you spot a snake before they do. You don't need to think about reacting; your limbic system identifies the threat, ties it to an emotion that signals threat, activates your fight or flight system, elevates your heart rate, and moves your arm to halt their progress – all before you and your neocortex can say "snake" and process feelings of fear.

In our primitive days of running from saber-toothed tigers, this kept us alive. In our modern world of rush-hour traffic and constant dings of smartphone notifications, we need to keep it in check so it's not chronically in low-level activation mode.

When asked about emotions and their evolutionary role, Dr. Anderson replied:

"My stance is that emotions are evolutionary, as well as cultural adaptations. Some of our research has even shown how emotions change the expression on our faces to help us perceive in different ways that are appropriate for the situation (e.g., fear opens the eyes, and actually changes retinal sensitivity, while disgust narrows the eyes to help focus, like inspecting whether something is good to eat or spoiled). Then we learn these expressions are powerful tools to communicate our mental states, and then they are used by culture to read and influence other people's behaviors."

Anderson continued "For each emotion, there have been proposed functions, whether it is sadness (rest and recovery from injury), disgust (avoidance of disease), fear (avoidance of bodily harm) or joy (play, exploration of the environment to build new resources). These are powerful tools that are meant to prolong our lives. But each can also be rendered maladaptive, moving us away from the capacity for flourishing."

"Maladaptive" and "dysfunctional" are the keywords here. Emotions and feelings have proven to be biologically advantageous through evolution, but when they're not controlled, they can consciously (and more often subconsciously) guide our decisions, often away from our stated long-term goals, in favor of what is most pleasant in the current moment.

In the next chapter, we'll look at what happens

when the limbic system becomes overactive, how we can train our prefrontal cortex to "watch it" (increase awareness) and how we can drive more conscious thought in an effort to reduce dysfunctional or maladaptive behaviors in our decision-making processes.

Reflection Questions

- Could you move more? How? How could you move in a wider variety of ways?
- Do you notice certain activities that improve your mood and certain activities – or periods of inactivity – that negatively impact your mood and decision making? How can you do more (or less) or these?
- Physiology check-in questions to ask yourself throughout each day: What signals is your environment sending your body and mind right now? How are you sitting right now? What is the position of your skeleton right now?
- In what position do you spend the majority of your time? How many submissive vs. beneficial poses do you occupy in a day?
- What two or three action items can you keep up your sleeve to instantly change your state?

Activities

Motion Creates Emotion

Emotions are physiological states. Therefore, changing our physiology enables us to change our emotional state. Our first activity here is motion. This is perhaps the most obvious tool at our disposal to alter our physiology.

Just check social media for all the quotes about "the only workout you regret is the one you didn't do" or "you're just one workout away from a better mood."

There's no doubt: movement changes our physiology.

A simple twenty-minute walk has been shown via brain scanning technology to significantly increase neural activity. [21]

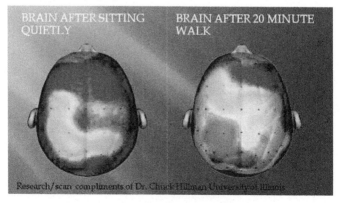

This doesn't even begin to mention the endorphins, the neurotransmitter cascade, or the long-term health benefits from moving.

I called this section "motion" rather than "training" or "workout" for a reason. All training is movement, but not all movement is training. For the purposes of this book, movement will suffice to positively impact our health and positively alter our states. And structured training would be an entire book (possibly several) itself.

Structured exercise is only one way to move. We can garden, hike, rock-climb, paddle a canoe or kayak, surf, paddle, practice martial arts, or lift weights, our options are limited only by our creativity – which, paradoxically, is stifled by inactivity. [22]

As you are likely already aware, our society has become increasingly sedentary. This has serious negative impacts on our brain function, our overall health, and our mood/happiness. [23]

Humans are made to move. Exactly how much more movement our ancestors got on a daily basis compared to our daily average is tough to quantify, but several recent studies have attempted to measure the miles walked and calories burned per day among modern societies that live by ancestral foraging methods, namely the Aché in Paraguay and the Kung of the Kalahari Desert in Southern Africa.

Here's what they found:

"If Americans want to be as active as the Kung, we would need to add the equivalent of 3.8 miles of walking to our daily activity. To be as active as the Aché, we would need to hike seven miles a day. Taking the average of these foragers, we would need to add 5.4 miles per day to move as much as our ancestors likely moved. Walking at an average pace of three miles per hour, this would take 1.8 hours, which is 3.6 times the thirty minutes recommended by the Surgeon General and other medical experts." [24, 25, 26]

To further support our need for movement, my recent collaborations with Dr. Andrew Hill, neuroscientist at Peak Brain Institute and gerontology professor at UCLA, have revealed that we need to accumulate 7,000 steps per day (5,000 at the bare minimum) for optimal health.

With the average stride length being 2.1-2.5 feet, it takes 2,000 steps to cover 1 mile. 5,000 steps would equal 2.5 miles, and 7,000 steps would be 3.5 miles per day, which translates to fifty to seventy minutes of stepping per day, again well above the Surgeon General's recommended thirty minutes per day. [27]

According to a Mayo Clinic study, only 46.5% of American are "sufficiently active" — a threshold that only requires 150 minutes of activity per week. An extra calculation of these numbers shows that thirty minutes per day over seven days (our woefully inadequate "daily thirty") totals 210 minutes of activity per week. So this 150 minutes per week (2.5 hours) falls short of that paltry standard, yet fewer than half of American achieve it. [28]

Our ancestors did that daily. That makes our "standard" for healthy activity seven times less than what our biology is used to. Not exactly a recipe for optimal biological and physiological function. No wonder we're obese, inflamed, depressed, and sick.

When we hear the advice to move for thirty minutes a day, or see wearable fitness trackers urging us to take 10,000 steps a day, we need to realize that these are minimum thresholds that merely get us to the level of no longer at "increased risk" to die young or develop mood disorders. Achieving these standards does not mean we have optimal mental or physical health. Rather, it means we're above the threshold for sucking.

The benefits of movement and physical activity are so vast that they easily expand beyond the scope of this book, so as a movement-obsessed performance specialist, it's hard for me to draw the line on what to include and what to leave out.

That said, indulge me as I expand this movement section a bit to include the following reasons we should prioritize frequent (not just daily, but throughout each day) movement.

As Dan Buettner discovered when writing *The Blue Zones*, his groundbreaking expose on geographical areas and cultures with the highest concentrations of centenarians, increased activity is related to increased life expectancy. Buettner writes:

"Our team found that people [in these places] are nudged into physical activity every 20 minutes or so. [...] They're walking to their friend's house. They're going down to the garden. They're kneading bread with their hands. It's natural movement. It's something they don't have to think about. It's not something that requires discipline." [29]

Movement has been linked to increased brain activity, fighting off depression, mood disorders, and maladaptive neurotransmitter function, increased creativity, and increased production of the brain-derived neurotrophic factor (BDNF) and nerve growth factor (NGF).

As we mentioned in the lead of this insert, a twenty-minute walk dramatically increases neural activity in the brain, as shown on brain imaging. Stanford researchers have shown that a ten to fifteen-minute activity boosts creativity, although they're not yet sure why or how this happens. [21,30]

A 2012 review of studies investigating the impact of movement or exercise on mood disorders concluded that regular movement increases perceived quality of life and is a "viable treatment for depression." [31]

The Duke SMILE (Standard Medical Intervention versus Long-term Exercise) study compared the effectiveness of traditional depression medications versus exercise as treatments for depression over sixteen weeks and found that exercise was equally effective at reducing depression symptoms. Even more encouraging was the finding that those in the exercise group were "50% less likely to be depressed at their ten-month assessment compared to non-exercisers." [32]

Harvard psychiatrist and author of *Spark: The Revolutionary New Science of Exercise and the Brain* John Ratey is a big proponent of exercise for its ability to enhance BDNF production. Ratey calls BDNF "Miracle-Gro" for the brain, as it literally signals the brain to create new nerve cells through a process known as neurogenesis. [33]

Sprints have been shown to increase BDNF, as

have other modalities of High Intensity Training (HIT). A 2011 study found that plasma levels of BDNF was significantly higher in Brazilian sprinters than it was in sedentary populations. As far back as 2006, researchers found that vocabulary learning was 20% faster after high-intensity sprints due to increased BDNF and acute increases in dopamine (pattern recognition). As few as three all-out sprints separated by two minutes rest are all you need to reap these benefits. [34, 35, 36] NGF is also positively impacted by movement. [37]

Don't go crazy with the sprints. Too much sprinting drains the nervous system, lowering heart rate variability (HRV) and actually decreasing BDNF. One or two sessions per week is likely optimal.

Ancestral health leader Chris Kresser points out that too much "cardio" increases inflammation (the root cause of all chronic disease) and oxidative damage, suppresses the immune system, decreases fat metabolism, disrupts cortisol levels and can lead to neurodegeneration. [38]

Rather than become a "cardio bunny" Chris echoes the longevity boosting advice from Dan Buettner: move frequently at a slow pace. He also advises occasionally lifting heavy, throwing, and sprinting. I agree.

Overdone, exercise can throw our body into defense mode, activating sympathetic responses that drain vagal tone (decrease HRV). As Kresser pointed out, this can lead to an incredibly potent cocktail of HRV destruction. Done correctly however, movement can reduce inflammation, increase HRV and emotional resiliency, make us smarter, happier, and help us live longer. [39]

How can you find the sweet spot? Do enough to stimulate performance progress, but not so much that life outside the gym suffers.

As a self-professed recovering fitness addict, I've been there. I've been the guy who defined himself by my body fat percentage and my weight room maxes. I've evolved and realized that wasn't going to be my legacy. I now focus those efforts on creating a lasting legacy, helping millions and showing others the path that leads so many of us to fulfillment, satisfaction, and dare I say...happiness! My final advice on exercise is that, like diet, it should enhance the rest of your life, not dictate it.

Again, this movement does not need to be "formal activity" – just *move*!

Moving More

Some ways to incorporate more movement into your day include stand-up desks, morning walks, parking at the back of any parking lot, taking public transportation or riding a bike instead of driving, and always taking the stairs instead of escalators or elevators.

When forced to sit for long periods of time, change positions often, use the Pomodoro Technique or set other timers to remind yourself to get up, take breaks and use those breaks to stretch, do your favorite yoga poses, practice tai chi or qi gong, do pushups, squats or mobility work – literally anything to lubricate your joints and increase lymph and blood circulation.

If you have to be on the phone, wear a headset and walk while you're talking. Be like Steve Jobs and make your meetings walking meetings to increase movement, spark creativity, and spend more time in nature all at once. Aristotle, Freud, Charles Dickens, and Harry Truman are also known to have made the walking meeting a staple in their schedule.

Movement-Based State-Hacking

Remember: motion creates emotion. At any time in your life, you can instantly change your physiological state with a set of thirty burpees. At work, you can break up the monotony of the day with a balance board or do squats or pushups every thirty minutes at your desk, or starting your day with a brisk walk in the early-morning sunshine (which also increases dopamine and serotonin via retinal exposure to sunlight).

Look for ways to "stack" physiological hacks to increase your return for investment. Another example would be to go surfing with some buddies. You're getting movement, play, camaraderie (oxytocin), time in nature, and grounding (direct skin contact with the earth/water).

Think of movement as your physiological gateway. You can use it to clear your head, think, process information, take a break, get space, increase brain activity, or create momentum that helps you overcome "writer's block" or the resistance of inertia to start whatever project you're delaying.

Motion creates (better) emotion.

Posture, aka "Don't Sit Like a Douche"

Posture is a powerful driver of physiology. Don't destroy your physiology by sitting or standing like a douche. – Aaron Alexander

We move less than ever before. We've established that fact. What's alarming is that the average American now sits 7.7 hours a day, and the CDC has started to say, "Sitting is the new smoking." [40, 41]

According to the CDC, chronic diseases are the number one threat to public health (not infectious disease), and the leading cause is, believe it or not, chronic sitting. [40, 42]

85% of America's workforce is paid to sit at a desk all day. [43] It's not just the fact that we sit; it's *how* we're sitting that is really crushing our physiology.

Posture can impact our physiology, feelings and emotions. In as little as two minutes, posture can change both our sex hormones and our stress hormones.

A (controversial) Harvard University study found that holding specific postures – known as power poses – increased testosterone by up to 20% in a mere two minutes. [44]

These stress-lowering "power poses" were the subject of Amy Cuddy's 2012 TED talk that has been viewed more than 44 million times. Her research has been challenged, and several attempts to recreate the results have failed. Since Cuddy's research has been questioned, let's do our own experiment, from my friend and movement expert Aaron Alexander. [45]

Aaron asks people to stand and pretend to be sad. Go ahead and try this one for yourself. Assume the position of a sad person.

You probably have a "dead weight" stance, with a rounded back, slouched and rounded forward shoulders, with your chin and eyes down. At this point, Aaron likes to hand the person a smart phone and *boom*, they're in perfect position to hold it at their sternum and disappear in to the world of social media. Something to think about.

Whether Amy Cuddy's power poses actually change testosterone levels, our position in space and the alignment (or misalignment) of our skeleton certainly impact our mood, thoughts, and decisions. Consider the way we sit – submissively hunched with shoulders rounded forward, head down, buried in a book or, more likely, an electronic device, serving at the beckon call of incoming messages, notifications, and email demands.

There, our spines are compromised; our glutes and hamstrings – the muscles that enable us to be bipedal primates – are nearly switched off, and our breathing is forced into our lungs with elevated traps and shoulders, all of which signaling our body that we're at capacity, a stressed state, leading to increased production of stress hormones.

Breathing into the lungs instead of the diaphragm is a signal to your body that you're at capacity. This is what you do when you're "out of breath." It sends signal to the body that we're at capacity or in trouble and triggers the stress response. This is good if we're running away from saber-toothed tigers, but it's not so good if it's happening 24/7 due to flawed posture and breathing patterns.

As you'll learn in Chapter 4, these faulty breathing patterns tell our vagus nerve that we're in a stressful situation, and sympathetic fight or flight is activated. It's this chronic stress stimulus, chronic information overload, chronic inactivity that crushes our health, mood, and feelings.

Here's another posture check: are your traps elevating your shoulders right now? As you read (or hear) this, can you lower your shoulders? Try to depress them (or think of extending your neck) so that your shoulders and ears move further away from each other. The more distance you're able to cover doing this, the more hunched over you were.

When you're freezing cold, what do you do? You clench everything, drawing your body tighter – especially your shoulders toward your ears. This is another sign of distress. Again, potentially useful when spending a rare night outside in brutal conditions, but harmful to modern human physiology when it becomes the default posture.

We can positively impact our breathing, physiology, and mood by simply increasing our awareness of our shoulder and spinal positions at all times.

When I sit, or even when I work at a stand-up desk, I constantly ask myself, "What does my skeleton look like right now?"

The postures that are bad for our health do not reflect ideal skeletal positions. I find this to be an easy, useful strategy to be more mindful of my posture and keep my skeleton aligned.

As for breathing, the more you focus on breath, the more your positive habits become routine. It's like anything else – intentional, focused repetitions lead to positive habits that become unconscious competence.

See the "Tools" section at the end of Chapter 7 for specific breathing strategies.

All the physiology gurus agree: move frequently

and avoid holding the same position for extended periods of time. The CDC recently released a poster showing that getting up and moving every hour reduces breast cancer risk by 21% and lowers diabetes risk by 30%. The risk of hypertension is reduced by 50%. So skip the fidget spinner and move your body; it's better for your health, happiness, and brain function. [46]

If you must sit for long periods, take the CDC's advice get up every thirty minutes and do one of the following as a pattern disrupt. If you must sit for several hours at a time, try to cycle through these to protect your skeletomuscular health, physiology, mental health, and productivity. Start simple. Get up and move. Stretch. Nothing formal, just move what feels tightest or most restricted. Do circles and eights with every joint: neck, arms (at the shoulder), elbows, wrists, ankles, knees, and hips.

Do some of your favorite stretches or yoga poses. Couple this with pictures of nature (shown to increase mood and brain function) or, if you can, go outside and do it. Walk. Get closer to those 10,000 steps every day and get some sunlight (bonus points if you take off your shoes and connect with the earth for some grounding).

Focus on your breathing. Are you belly-breathing? Or are your traps activated and shoulders elevated?

I like to throw in some crawling and free moving just to break the linear patterns. Crawling serves as a neurological reset and can be as simple as baby crawls, bear crawls, or crab walks. You can move from bear crawls to crab walks seamlessly in what looks like a rookie breakdance move for some "free moving." Have fun with it. Add music. Now you're smiling, moving, listening to music you like, and improving posture – all of which positively impact emotions, feelings, and thoughts. For more on crawling, check out Tim Anderson's *Original Strength*.

Here's a guided five-minute recharge I put together for our House of Strength members a few years ago. I creatively called it the HOS Recharge.

Perform these three moves as a circuit and repeat the circuit for three rounds 1-2 times each work day.

1. Crucifix stretch x 10 seconds (Hold your arms out by your side so you resemble the shape of person on a crucifix, turn your palms to the ceiling, drive your thumbs backward, squeeze and hold for 10 seconds.)
2. The Butt Flex x 10 seconds (Stand up, spread your feet, try to grip the floor with your toes and squeeze your glutes as hard as you can for 10 seconds.)
3. RKC Plank x 10 second (Hold a plank position, with your elbows on the floor. Imagine driving your elbows down and pulling your feet up. Neither will move, but your intent should create maximum tension throughout your body. Hold this for 10 seconds.)

Finally, keep some exercise bands, lacrosse or tennis balls at your desk for assisted stretching, mobility drills or trigger point relief.

Don't overlook the value of recovery and the flushing of metabolic waste. Research has shown that the removal of mitochondrial waste product can improve carbohydrate metabolism and reduce the risk of diabetes as well as other diseases of chronic inflammation. Each of the pattern disrupts listed here is a method of recovery and serves as a posture check that can help us attenuate the harm caused by sitting 7.7 hours a day and staring at screens for an average of ten hours a day. [47]

If you find your days, or your life, getting out of alignment with your stated goals/objectives, look no further than the feelings created by your common postures. What position is your body in most of the time?

Remember posture impacts physiology, and physiology drives emotions, feelings, and decisions. Posture is an important domino that has profound downstream impacts on health, strength, brain function, and happiness. We're always in a posture; make sure you're consciously choosing yours.

Positive, happy, successful people in great moods rarely sit like a douche.

CHAPTER 3:
Running the Show – The Battle Between Our Prefrontal Cortex and Limbic System

Marshmallows and the Ulysses Pact

How you do anything is how you do everything.
– Buddhist saying

Imagine sitting in an empty interrogation room. It's a cold, featureless room, with only a metal chair and table. In front of your table is a two-way window where you know people on the other side are watching you.

An emotionless man in a white lab coat hands you a marshmallow and says, "If you can delay eating this until I return, I will give you a second marshmallow."

What do you do? How long would you last before eating the marshmallow?

This is the premise of the famous and oft-cited Stanford Marshmallow Experiment, where kids are given a marshmallow and left alone in a room. The researchers tell the kids that if they can hold off on eating the marshmallow until the researchers return, they will be given two marshmallows. [48]

What the kids don't know is that the researchers aren't coming back. Eventually, the kids give in and eat the marshmallow.

Interestingly, the researchers continued to follow those kids throughout their lives, finding that the kids who delayed eating the marshmallow the longest performed the best in school, went to better colleges, got better jobs, and generally led more "successful" lives.

Why? The single marshmallow represents instant gratification. Those who delayed eating the marshmallow the longest demonstrated that they had tools or techniques to help themselves delay gratification.

Those same strategies served them well throughout life, leading researchers to conclude that was the reason for their higher performances in school, work lives, and life in general.

The first time I heard this study being explained – even before I heard the results and findings – I was envisioning myself in that position, wondering what I would do. I don't eat sugar. I don't eat junk, candy or marshmallows, so in my mind, I imagined all the different ways I could destroy the marshmallow so I could not eat it, from tearing it up, to smashing it onto the two-way mirror where researchers were watching me, crushing it under the chair, or smashing it flat to stuff it under the crack of the door.

I also saw myself doing this immediately, as in the very second that marshmallow was presented to me. This action taken at my moment of highest alignment with my desired outcome would ensure that no matter how weak my future self became, there would be no marshmallow to eat.

As I've interviewed neuroscientists and high performers for this book, I discovered that behavioral scientists call this action a Ulysses Pact.

This pact between present and future self is named after the literary character Ulysses (Odysseus, in the original Greek work), who craftily fashioned himself to the mast of his ship in order to immobilize himself, so he could not be tempted by the sirens' songs that lured ships into rocky waters and ultimately sank them, killing everyone on board.

To further buoy his chances of survival, Ulysses had his crew fill their ears with wax, so they would not hear the sirens' song. He directed them to ignore his predictable temporary madness that the songs would incite, and only to untie him after they had passed the island. As extra incentive for his men not to untie him, he promised to kill any man who betrayed his directives.

Ulysses knew of the coming temptation, he knew his limits, and he made a pact between his current and future selves. He rigged the game to ensure success for his future self.

This is a strategy that high performers use throughout their lives, and it is one that can surely help us delay instant gratification in the pursuit of our highest goals, no matter how good the marshmallows look in the immediate present.

Awareness

Awareness creates choice. – Bill Harris

So far, we have explored the neuroscience that explains how our subconscious beliefs and operating systems assign meaning to our physiological states, which in turn creates the emotions and feelings that drive 95% of our decisions.

In this section, we'll reverse-engineer the decision-making process, working backwards from our desired result to identify the actions we can control to cultivate the physiological environments that facilitate high performance and enable us to make the choices that are congruent with our goals we have for our lives, no matter what the internal (or external) circumstances may be.

In other words, this is where we learn how to create an unbeatable mind that provides us the fortitude, resiliency, willpower, or mental toughness to see things through, overcome adversity, and get the results we want.

Back to our original question, if we know where we want to be, why is it so damn hard to get there? What prevents us from becoming that person? Why are successful, high-performing people the exception, rather than the norm?

As we've seen, our beliefs and operating systems – the subconscious programming we run on, the narratives we tell ourselves, the resulting emotions, feelings, thoughts and decisions that they drive – are unwittingly leading us astray.

When these transient emotional states take the wheel, they guide us in the direction of impulse and instant gratification, not necessarily the original plotted destination. In these situations, we're like a raft set adrift on the vast ocean, subject to the whims of the tides and winds. We are far from in control of our course.

It'd be like stopping at every exit on the highway because of a compelling billboard or shiny object when you know your destination awaits you 300 miles down the road.

It stands to reason that increased awareness of

these internal battles will afford us better choices, which can drive better decisions, leading to better actions, and ultimately, outcomes that are more in line with our desired goals. Start acting as if you already are that version of yourself that you want to be.

In this section, we'll explore how heightened awareness and elevated consciousness reduce blind spots, show us choices we may not otherwise see, and facilitate better decision-making. We'll also explore something scientists call choice architecture – a method of making deals with our future selves, like Ulysses did, to limit potential choices and create better defaults for ourselves.

To understand how we can heighten our awareness, create more (and better) choices and ultimately control our thoughts and decisions for more favorable outcomes, we need to meet our limbic system and our prefrontal cortex.

Limbic System, Prefrontal Cortex, and the Evolution of the Human Brain

The brain's failure to regulate itself is the source of all human struggle. – Bill Harris

So far, we have looked at the difference between feelings and emotions, and how our feelings drive our decisions. Now, we're going to explore how our decision-making process and our emotional resiliency are linked to which of the several internal brain systems that are in control at any given time.

Like parallel universes, we have several competing operating systems in our brain that are constantly negotiating with each other, locked in a perpetual game of tug of war. The main systems that we'll highlight here are our limbic system and our prefrontal cortex. You may recognize these from the triune brain theory presented in the last chapter.

Enter Bill Harris, founder of Centerpointe Research Institute, specialist in neurotechnologies for human change, evolution, and healing. When I interviewed Bill a few years ago, he did a terrific job explaining how the limbic system and the prefrontal cortex are at constant interplay inside our minds.

Think of it like having a Zen monk and a hormone-driven, rebellious teenager fighting to control your mind. The impulsive, ego-driven teenager in this analogy is the limbic system, and it is responsible for our propensity to be tempted by mythical sirens, blow off healthy appointments (workouts, meditation time, salads at business dinners), ditch our plans to start that book or business we're been thinking about, spend money we know we shouldn't be spending, or any other dumb decision we know is driven by instant gratification-seeking behaviors that provide nothing more than short-term satisfaction.

Bill brilliantly details the difference in both speed and predisposition of these systems, reminding us that it is the limbic system that fires when we walk by the bakery section at Whole Foods and immediately react with: "Oh, those cookies look amazing!"

After that initial reaction, the prefrontal cortex kicks in to remind us that while those cookies look tempting, they may not be our best choice – something called cognitive control.

As we detailed in Chapter 1, the limbic system can only focus on the "here and now." It is a compulsive, ego and dopamine-driven system. It is linked to our sympathetic nervous system, commonly known as the fight or flight response.

When this fight or flight response is activated, blood flows away from the brain to our extremities for mobilization in response to the perceived threat. This hardwired defense mechanism actively diverts metabolic resources away from our brains in these situations, literally making it harder for us to execute higher-level mental functions.

The limbic system is incapable of abstract thought and cannot look at long-term consequences. Now, the limbic system is not all bad. Nothing is inherently good or bad, as we must apply goal and context to the assessment of everything. It's the balance that we must pay attention to.

On the positive side, the limbic system is linked to love, creativity, and those survival responses mentioned earlier. The limbic system is beneficial, even life-saving in certain scenarios. But if we let it, the limbic system will overstep its bounds and begin to control our decisions outside of its biologically intended domain.

The problem arises when our modern life has us living in chronic low-grade fight or flight. That's why we freak out if someone cuts us off in traffic, and only much later do we realize the downstream impact this has on the rest of our day. (Yes, I'm talking to you, Mr. Posted on Social Media about some idiot in traffic).

The bigger issue is that, like any neurological process – shooting a free throw, throwing a football, pressing a barbell, or anything most people might confuse for muscle memory (it's actually neurological patterning; muscles don't have the capacity for memory) – repetition creates habit.

The more we use this system, the stronger it gets, and the more it becomes our "default" pathway. And like any other bad habit, this makes it harder to break and means we need even more "good reps" with the desired behavior to pattern the new pathway.

Think of this as your brain trying to use the interstate instead of an unpaved dirt road. It wants to use the familiar, least resistant pathway. We have to consciously use that dirt road option until it becomes as worn in as the highway, eventually becoming *the* way.

Running and biomechanics expert Scott Dolly says this re-patterning can take between 3,000-30,000 repetitions in the new pattern to make it *the* way. The less conditioned the human, the more reps this will take. If we applied this to running with a new foot strike, a trained athlete may only need 3,000 repetitions to change their neural patterning, while a non-trained, sedentary "couch potato" with little kinesthetic awareness may require closer to 30,000 repetitions (or more).

We must be prepared to exercise patience as we work to rewire our brains. It won't happen overnight, and the more we focus on awareness, the better our transition will go.

Here's the good news. The prefrontal cortex is involved in pattern recognition, learning, and executive decisions, and it supervises the limbic system, so you're actually training it to become good at its natural role.

But we will have to train it because, as Columbia University professor and author of *Computational Psychiatry: A Systems Biology Approach to the Epigenetics of Mental Disorders* Rodrick Wallace reminds us, "Emotions are self-regulatory responses that allow efficient coordination of the organism for goal-oriented behaviors." [49]

When they're working properly, that is. This system of self-regulation is the very reason every neuroscientist has to qualify their statements with some variation of "when functioning correctly," or "when not maladapted."

And as Bill Harris reminds us at the lead of this section, "The brain's failure to regulate itself is the source of all human struggle."

Self-regulation has its drawbacks, and as we've seen, the limbic system is one that needs to be chaperoned closely. While this concept is simple in theory, it is not easily implemented, and even harder to master. Look at this recent study showing just how dysfunctional we are at self-regulating our Facebook usage:

"The study confirmed that insufficient self-control is a predictor of Facebook addiction. This is in accordance with the general trend presented in the subject literature, stressing that people with good self-control cope better in life, can hold their temper, do not abuse alcohol, keep secrets, and save money [50].

Also, some findings indicate that people with a high level of self-control can use the Internet in a healthy way [51]. Our results suggest that people who have insufficient self-control have problems with dysfunctional Facebook use."

Breaking news: "People with a high level of self-control can use the Internet in a healthy way." As this study confirms, people with better control over their limbic system generally make better choices across domains in life.

Conversely, people with insufficient self-control, aka limbic system running amok, exhibit compulsive behaviors like "dysfunctional Facebook use."

Perhaps more significant in our pursuit with this book was the following paragraph in the researcher's conclusion:

"Our study showed that low level of failure-related action orientation is a predictor of Facebook intrusion. In general, people with prevailing state orientation experience more strongly engage in thoughts related to unpleasant situations involving failure, focus on negative emotions, cannot easily overcome these states, and have problems with the regulation of emotions, thoughts, and behaviors; moreover, they do not look for a solution." [50]

In other words, people who let their current states (feelings) guide their decisions, "more strongly engage in thoughts related to unpleasant situations." F*ck your feelings!

This statement also supports the notion that self-regulating systems cannot objectively measure their impact on the goals of the greater system – the "team" in our sports and military analogies, or in the real-world, our stated life's goals.

Paraphrasing the researchers, people with

dysfunctional limbic system not only engage in negative "emotions, thoughts and behaviors," but also, "they do not look for solutions."

Either they're unaware of this dysfunction or worse, they don't care. Whether our limbic systems are functional or dysfunctional, you and I certainly care; if you didn't care, you wouldn't be reading this book.

High performers don't become high performers by accident. Evolved, enlightened people do not elevate their consciousness by accident.

An examination of the current mental health of the American population and our economy further buries the notion that emotional self-regulation can be in line with our goals for becoming the happiest, healthiest, strongest, and highest expression of self that we can be.

The US currently boasts a $210 billion-dollar mental health industry (conservatively estimated), and a $4 trillion-dollar "altered-states economy," as explored in the book *Stealing Fire* written by Jamie Wheal and Steven Kotler. [52, 53]

Stealing Fire explores flow states, ecstasis (from ecstasy), and our growing obsession with gambling, alcohol, porn, and other mind-altering substances used regularly to (quite literally) get out of our own heads.

One thing to note about the work of Kotler and Wheal: their previous book *The Rise of Superman* discusses the value of flow states for optimal performance. In the neuroscience of flow, the brain enters a state of *hypo*frontality, meaning the prefrontal cortex significantly reduces its activity. This is good *during* performance and should not be confused with the states of awareness we've previously mentioned. When performing – be it surfing, rock climbing, or playing an instrument – we want to rely on autopilot systems. If Tiger Woods is thinking about his golf swing on the course during a tournament, he's done. Likewise, if an extreme sports athlete is thinking about the mechanics of their sport while performing, they're done (or dead).

Looking at these numbers and reflecting on our own personal experiences, it doesn't take long to realize that this self-regulating system is not exactly supporting our goals of emotional evolution and self-optimization. Our biology is wired to look for threats in order to help us survive. We're wired to react quickly. This neocortex or prefrontal cortex is new, and we're still learning how to put it in charge.

We also have the powers of general observation that show us we live in a world of pain, suffering, and fear, all of which support the notion that these self-regulating systems can run amok if we don't keep them in check.

But why? Why are we seeing so much dysfunction? The short answer is that we're still the same species we were 2000 years ago. And 2000 years ago, we did not have electricity, climate-controlled environments, cars, planes, the Internet or smartphones, to name a few of the significant changes to the way we live.

We spend less time outdoors and more time sitting in air-conditioned (or heated) spaces lit by artificial light. This scenario alone disrupts our normal movement, posture, socialization, and light exposure, all of which impact our physiology in major ways –

and as you now know, physiology dictates feelings.

Compared to 2000 years ago, some of us – especially those in major cities – encounter more people in a single day than we would in an entire lifetime, yet we spend less time engaged in meaningful face-to-face interactions. We're consuming exponentially more information on a daily basis than ever before, and we're constantly bombarded with unrealistic standards in marketing and social media.

These are glimpses of our new reality, and they're all too common issues that lead to dysfunction with the limbic system and our lizard brain. Our brains simply haven't evolved and adapted to our new reality. It makes perfect sense that we're struggling to adjust.

"Well-adjusted," anyone? Anyone?

The blissful, peaceful, happy, safe and utopian state we wish was our default setting can be attained through practice, but it's not our default setting. Remember: we're wired to be on high alert, to constantly look for threats and danger in order to stay alive.

Here in the first world, we have the luxury of roofs, beds, and food security. Our survival needs are met, but our primitively wired brains will run off unchecked if we allow them to do so.

To paraphrase Tony Robbins, we can always find always something wrong if we view our world in that manner. And we're wired to look for and spot those potential threats. We have to train our minds to work for us, not against us.

How can we do that?

The big question now becomes, how do we keep the limbic system in check? And how can we enhance our prefrontal cortex, make it stronger, and help it prevail over the faster, more compulsive limbic system?

We can use the following four-pronged approach.

We are wired to see what's wrong – that's what kept us alive and helped us evolve as species. We have to:

- be aware of this;
- be vigilant and watch our brains;
- do the work to rewire our brains;
- develop our prefrontal cortex so it becomes our prevailing system.

Awareness over a self-regulation system is crucial in our efforts to make it less self-regulating. That lack of oversight leads to, at best, unchecked belief systems, at worst, dysfunctional emotions and decision-making.

As Harris points out with the statement "awareness creates choice," increased awareness is the first step in being able to direct our conscious thoughts to facilitate better decisions and actions.

Consider this 2008 study that followed 101 stock investors, each given $10,000 to invest. The study, designed to examine the link between affective experience and decision-making performance, had all 101 investors rate their feelings for twenty consecutive days of trading. [54]

What they found was interesting: the traders who were most consciously aware of their states of mind were also the individuals who best demonstrated the ability to make rational decisions. The researchers wrote:

"Individuals who were better able to identify and distinguish among their current feelings achieved higher decision-making performance via their enhanced ability to control the possible biases induced by those feelings. [...] The advice to 'Ignore your emotions,' appears, in view of our results, not to be the right answer for effective regulation of feelings and their influence on decision-making. Instead, the results suggest exactly the opposite: individuals who better understood what was going on with their feelings during decision-making and thus reported them in a more specific and differentiated fashion were more successful in regulating the feelings' influence on decision-making and, as a result, achieved higher investment returns."

As we can see, high performance decision-making is less about ignoring our fears or other emotions and more about acknowledging them, then refusing to allow them to cause emotionally driven deviations from our planned course.

As Dr. Harris notes, this increased awareness of mental states affords us the ability to have more conscious control over what may otherwise be a subconscious-driven decision.

Increased awareness keeps our conscious mind in the driver's seat, not our emotional, rash, impulsive limbic system.

As neuroscientist David Eagleman, bestselling author of *Incognito: The Secret Lives of Our Brains* wrote, these subconscious decisions tend to run on autopilot (by definition, without our conscious awareness), and these systems are generally a good thing as they allow us to focus on higher-level tasks instead of mundane, menial processes. [55]

The problems arise when the belief systems and narratives that guide these subconscious activities are running a flawed software. While beneficial, even essential, these systems must have some conscious, mindful oversight on our part to ensure that they're steering us in the direction we want to go.

First, the underlying belief systems on which these subconscious decisions are based need to be periodically audited, as Vishen Lakhiani explains in his book *Code of The Extraordinary Mind*.

Vishen calls them "Brules," or Bullshit Rules that we accept and allow to guide our lives without consciously assessing them. An example used in his book is the institution of religion. Vishen was raised in Malaysia by parents who did not eat beef for religious reasons. At some point in his own life, he reassessed this belief system and created his own values around this construct. I'm greatly oversimplifying his concept for sake of brevity, but I think this is sufficient for you to understand how this plays into what we're talking about with awareness and how our belief systems shape our decisions. [56]

Second, we must remember that this system is a self-regulating system susceptible to dysfunction. Periodically checking in with our values and realigning our subconscious with our conscious mind will help ensure that we stay on the right track.

It is not a "set it and forget it" system. We must constantly seek knowledge and growth, challenge our beliefs, be open to shift our paradigms, and make sure we're viewing the world and ourselves through the lenses of our core values.

Better awareness of situations affords us more and better choices. These choices, as Dr. Harris describes, impact us in four areas:

- How we feel
- How we behave
- What we attract and what we're attracted to
- What meaning we assign to things.

Of crucial importance is that most of these processes are normally functioning on autopilot, aka processed subconsciously. This is great if we've consciously (with awareness) laid the foundation for the system that handles those processes. Most of us have not done this work.

Remember: the meaning that we assign to things is inextricably linked via emotion to a fact that will later dictate feelings and drive decisions. Recall that things, events, and occurrences – nothing has inherent meaning. Things only mean that which we assign to them.

The aforementioned neuroscientist David Eagleman offers similar advice, writing that the left hemisphere of our brain is always looking for justifications and searching for meanings. Trouble arises when it can't find meaning and makes mistakes.

This is further reason to increase our awareness, elevate our emotional intelligence, and avoid sabotage from our subconscious.

We have no choice in whether or not our bodies

and brains will do this. Our choice in the matter is whether or not we choose to consciously direct it and guide it through mindful work on defining our own values, belief systems, and decision-making procedures.

This is one area where our high performers differ from "normal" people. They understood the need to, in some form or fashion (and not always the healthiest way), install systems that increase their odds of success.

Are your actions congruent with your stated goals?

This is a question I ask myself, my wife (carefully), and my clients all the time.

As I investigated the decision-making process and set out to write this book, I realized that this question or mantra was making one massive and dangerous assumption: that all choices were being made consciously.

As we've seen, they're not.

Consider your choice to own or rent the place in which you live. 99% of the people who read this book either rent or own their home. James Altucher, paradigm-shifting author, podcaster, and entrepreneur, does not rent or own. He constantly lives in one Airbnb after the next. Technically this is renting, but his paradigm is completely outside of the belief system that is unconsciously cemented into many of our minds.

Want another example of an unconscious choice you make every day?

How about eating cereal with milk and orange juice for breakfast? Or the idea that breakfast is the most important meal of the day. Or that you should sleep in a bed. In a bedroom. Or leaving the house with clothes on. By choosing *once* to not live in a nudist colony, we unconsciously agree to the societal norm that we will not leave our house naked.

I'll stop there before things get crazy. The point is, we all make decisions every day without conscious thought.

While not all decisions are made consciously, our values are defined consciously. The more aware we are of our values, the easier it is to make decisions – both conscious and subconscious – that are aligned with those values.

If we're to continuously, relentlessly make forward progress on our dreams, to truly master our fate and direct our lives, we need to increase our consciousness (wake up!) and bring an elevated awareness to every action we enter into.

The first step to ensure that our actions are always conscious is to be aware. This is why awareness is the first section of the "better decision-making" portion of this book.

Show up to everything we do with 100% presence and be intentional about every conversation, step, bite, and interaction. Being intentional isn't something out of our reach. It's more about reducing distractions, i.e., not using your smartphone while talking to someone in person.

As the Zen Proverb says,

> When walking, walk. When eating, eat.

From this position of increased intention and presence, we can maintain that 50,000-foot, objective view of ourselves. We can be vigilant and ensure that we're making choices that are in line with our stated goals.

Eagleman proposes an index of consciousness in which consciousness levels depend on abilities to mediate our brain's conflicting systems.

In his words, "One's degree of consciousness parallels intellectual flexibility."

Expanding our awareness and consciousness gives us more control over these underlying operating systems that drive our feelings and behaviors.

Small changes in brain chemistry and patterning can lead to big changes in behaviors. Awareness is the first step in our efforts to be better decision-makers.

The degree to which we are conscious (or aware) parallels our intellectual flexibility. Awareness creates choice.

Increasing Bandwidth

The next step is to focus on increasing bandwidth, or as Bill Harris describes it, raising the threshold at which the limbic system would kick in.

Research has shown that both the limbic system and the prefrontal cortex shrink and grow with use, so the more we train it, the stronger it gets, and the easier it becomes for us to control our mental processes. This is why meditation works. [57]

Meditation enhances the prefrontal cortex and shrinks the limbic system. [58] It is one of the tools we can use to help us train the brain to use the proper pathways for higher-level, conscious thought.

This is why the saying "the harder meditation is for you, the more you need it" is so profoundly on point. The same goes for the one about people who don't have the time for meditation – they're the ones who need it most.

Simply put, our bandwidth is limited. We all have a certain capacity for the amount of "stuff" we can deal with before we hit our breaking point. This threshold, as Harris describes it, is where the limbic system kicks in.

Stressful events shrink our bandwidth – hard workouts that drain our CNS, traffic, bosses, co-workers, screaming children, deadlines – these all occupy bandwidth, leaving us with little extra to deal with life's next adventure, or for our own pursuits in our precious spare time.

The prefrontal cortex has the ability to assert cognitive control in these situations, provided we have the emotional resiliency (the bandwidth) to process such higher-level thoughts.

How much cognitive control we have depends on the amount of bandwidth we have. Think of it like using the Internet at your house, where you also have limited bandwidth. If the TV downstairs is streaming Netflix, a computer upstairs is uploading a video to YouTube, and a cellphone is trying to play a video from YouTube, all are likely to suffer as the total bandwidth capacity is being taxed.

Meditation can increase bandwidth, as can yoga, breathe work, or anything that involves shutting down the sympathetic nervous system (fight or flight) and spending more time in the parasympathetic (rest and digest) state.

It's worth noting that supplements like L-taurine, L-theanine, and GABA (gamma-Aminobutyric acid) calm the limbic system. We'll talk more about GABA's importance in the chapter about the chemicals in your head.

In the next section, we look at how the vagus nerve plays a crucial role in the two-way street that connects our body and our brain, and how increasing something called vagal tone (or heart rate variability) may be synonymous with increasing the bandwidth we just discussed.

Reflection Questions

Your beliefs become your thoughts,
Your thoughts become your words,
Your words become your actions,
Your actions become your habits,
Your habits become your values,
Your values become your destiny.
   – Mahatma Gandhi

- Do you feel like you have shiny object syndrome?
- Are you letting your limbic system run the show?
- What are some scenarios where you could develop your prefrontal cortex's ability to control your limbic system?
- Which of your brain's operating systems is running the show when you make great choices? Which of your brain's operating systems is running the show when you make decisions that are not aligned with your long-term goals?

Awareness Creates Choice

How can you cultivate more awareness in your life? How can you be vigilant and more aware of the reasons behind your actions and choices? How can you examine your motivations and feelings?

Choice Architecture

Where are you setting yourself up for success? What "defaults" have you created to ensure this success? Where have you failed to do this? Where in your life can you use choice architecture or the Ulysses Pacts to create better defaults for yourself?

For example, a person looking to lose weight or increase movement could say, "I will always take the steps when presented the opportunity." Then, in the moment, choose the stairs over the escalator.

Activities

Choice Architecture

For our first activity, or tool that we can add to our optimization and self-mastery toolbox, we need to revisit Ulysses. The deal Ulysses made with his crew and his future self has become the model for an action that cognitive behaviorists call the Ulysses Pact. It's also known as "choice architecture."

As I've continued to research decision-making and talk to neuroscientists, it turns out, I've been onto something all this time.

For years, I've had a favorite saying that I relay to nutrition clients and gym members: "You can't eat what you don't have."

This advice is built on the fact it is impossible to prepare a healthy meal at home if you have not previously acquired said healthy foods. Literally, we have zero chance of sticking to our diet if we don't first acquire the right foods and fill our kitchens with them.

What I really love about this saying is that it holds equally true for poor food choices. If you never buy Oreos at the store, it significantly reduces your ability to eat foods that are not aligned with your stated goals.

Side note: if you're trying to cut down on poor food choices, here's a simple tip: stop f*cking buying them! Seriously, what do we think is going to happen when we bring those Oreos and ice cream home?

Back to my saying, it turns out that what I was coaching people to do is an example of this choice architecture. Choice architecture involves setting up better defaults so that when decision-making time comes, we're more likely to make a better choice, regardless of our feelings or emotions in that given moment.

In a 2003 paper, researchers posited that cafeterias should be organized in a way that the first foods seen by customers are the healthiest options. Such positioning, they hypothesized, would increase the chance that a) hunger and b) mindless consumption would prompt food choices more in line with stated goals to eat healthy. [59, 60]

It worked. The subjects in the study made significantly healthier choices simply due to the rearrangement of foods presented in the cafeteria line.

Here's another example from my life: I refuse to travel without healthy snacks. I know there will be situations when I'm hungry and won't have access to the quality food choices I prefer. Rather than concede to situations outside of my control and hand over responsibility for my own actions, I stay in control and can stay on track because my past self took some steps to ensure my present self has better options at hand. It's about awareness, preparation, strategy and commitment to values.

Second, the cafeteria study highlights how mindless most of us are as we move through life. Not just in food choices, although that is a major area of mindless choice in the US. But many of us may be supremely vigilant with food choices, yet extremely mindless with vehicle maintenance, clothing choices, or any of the thousands of other daily decisions we make subconsciously.

This is not inherently good or bad. There are areas of life where this can be used to our benefit, and there are areas of our life where this can unknowingly lead us astray. This is why awareness is the first step in becoming a better decision-maker.

Steve Jobs's decision to wear the same black turtleneck, jeans, and tennis shoes every day demonstrates the power of putting unimportant choices on subconscious pathways in order to protect our mental bandwidth for more important choices. The crucial element to note is that the original decision was made with heightened awareness of values (wear same thing every day, stop wasting brainpower deciding what to wear, focus on important stuff) and done in a way to make future Steve more effective and more aligned with his goals and values.

Some areas of our life, like our morning commute or our wardrobe, can be put on autopilot. Others, like our relationships, child-rearing, and our personal development, should not be running on these subconscious pathways.

Realize that we are emotional, irrational human beings wired to make choices based on our feelings. Then start looking at the default choices we've set up for ourselves in our life. As my mentor Paul Reddick told me back in 2012, "Your life is perfectly designed for the results you're currently getting."

I'll never forget that line. And when I realized I was watching twenty-four to thirty-six hours of TV a week (two hours each night, plus four or five hours a day on weekends), I realized I was spending a day or even a day and a half every week watching other people live their dreams instead of spending that time making my own dreams a reality.

Even on the low end, that one day per week came out to be 1/7 of my time, or 14% of my life. We immediately got rid of the TV. In the following twelve months, I built my gym, House of Strength, from zero members to a six-figure business.

Coincidence? I'm going to say no. Were there other factors and decisions involved? Absolutely. But many of those other factors involved setting up better default environments. I bought business and self-improvement books and left them in every room of my house, along with notepads. Guess what I started doing instead of watching TV?

The time to set up better defaults for ourselves is when we are most aware and most aligned with our values. I could have said, "Do this when you are most motivated," but motivation does not last. Values last. Alignment is not permanent, but it lasts longer than motivation and can be realigned easily.

Sticking with our "you can't eat what you don't have" example, do your grocery shopping at ideal times (like after a workout, when you're motivated to eat healthy), so you have better choices at home, do your meal prep, pack healthy snacks, and set yourself up for success.

Apply this to every area of your life where you

want to make a change — finances, time management, sleep, nutrition, workouts, business, or relationships.

Success is never an accident. It may not happen as expected, but few people achieve the lives of their dreams by accident.

Like my grandfather used to say, "Prior preparation prevents piss poor performance."

Creating Better Defaults

I've had the unique opportunity to learn from, befriend, work with, advise, and study the habits of nearly a dozen Navy SEALs over the last five years.

SEALs represent the epitome of mental toughness, and I love learning about their methods of honing elite human performance.

One of those SEALs was kind enough to answer a few questions about BUDS (Basic Underwater Demolition School) on the condition of anonymity. I asked him how he made it through BUDS, the six-month Navy SEALs selection process that is famous for its drop-out rate of 80% or higher.

His answer was simple. He made a conscious commitment before the first day that he was not leaving a failure. He would not ring the bell that BUDS candidates ring to signal they've given up. He eliminated quitting as a choice. His choices became die trying or pass. Once he removed quitting as an option, it left his mind, and he never thought about it again. Make no mistake: dying was an option, but quitting wasn't.

He created a different default option in his head before starting the journey, and in his lowest moments, his internal dialogue was drastically different than that of the 80% who quit. While the quitters' choices were a) continue to be miserable or b) quit and go get warm, our hero's choices were a) continue and pass or b) continue and die trying to realize his life's dream. Quite literally, his only choice was to keep going. When we look at that mental dialogue, it's really no surprise he made it through.

In case you're wondering, I had to ask: how many times did he think he was going die? More than once. But his lowest point, he told me, came during a drill that is no longer in use where he actually thought, "Either I'm going to die or I'm going to finish, but I won't quit." He kept his wits about him, kept treading water (literally), kept focusing on one breath then the next, and made it through – both the drill and the selection process.

It reminds me of a quote from Will Smith (another high performer by any measure of the term):

If you and I get on the treadmill, one of two things is going to happen: either you're going to get off first, or I'm going to die on that treadmill.

Successful people, whether they know it or not, are using choice architecture to manipulate their choices into more favorable outcomes, increasing their odds of adopting behaviors that are congruent with their long-term goals.

Do you need to risk death to win a bet on the treadmill or become an elite warrior? Probably not. But altering your *perceived* choices can have profound impacts on your results and your life.

There are two things over which we always have control: our attitude and our effort. I challenge you to first identify ways you are already using choice architecture to your advantage in your life, and then look for areas you can strengthen by incorporating this powerful behavior to influence better decisions in predictable situations of low willpower (they will come; it's our responsibility to ourselves to be prepared for them).

I used the word "willpower" intentionally, so that I can tell you that I don't believe in willpower. I believe in commitment to values. Fatigue and low mental bandwidth (poor vagal tone) can contribute to moments of "low willpower." If we examine those times in our lives where we felt willpower was lowest, we're likely to realize that fatigue and decreased vagal tone were factors. We're also likely to see some temptation that, like the marshmallow, promised instant gratification.

If you find yourself feeling low on willpower, shift your focus from the present circumstances and ask yourself how you might handle this situation if your spouse or children were with you. How might you handle this situation if you first had to write down your five core values? How might the ultimate version of yourself handle this situation? Act as if you were that version of yourself, not the one in this current physiology state. F*ck Your Feelings.

If we're truly committed to ourselves and our values, then we have infinite wells of willpower.

And we also have the following methods of creating better defaults for ourselves:

- If you want to get more steps per day, make it a "rule" that you park in the back of every single parking. No questions about it; parking in the back of the lot is just what you do. Always take the stairs, never the escalator or elevator. Make these default choices, and you will see a marked difference in how you act when presented with those options.
- If you want to cut out sweets, make it a rule that you don't eat any foods with added sugars. Now it's no longer a decision. There is no wiggle room – no room for rationalization in the moment.
- If you want to write a book, wake up early, block off one or two hours each morning, and write 1,000 words before you do anything else each day.

As you can see, it's not about willpower. It's about commitment to our core values and consciously designing our life to produce the results we want.

Focus on awareness, create choice, use this heightened alignment to create better default choices, and surround yourself with a community that celebrates you and encourages moonshots.

Instilling habits like these will directly generate progress in your life, and they'll lead to indirect progress as you train your mind to make the best decision for your goals – even when it's not the easiest or most convenient.

You will train your mind to think like the person

you *want* to be, regardless of how you feel in an any given moment. The more we do this, the more we actually become that person.

Sleep

Sleep is one of (if not *the*) single biggest factor in increasing our health, emotional bandwidth, vagal tone, and emotional resiliency. Simply put, inadequate quality or quantity of sleep results in measurable decreases in HRV (heart rate variability), which leads to reduced emotional resiliency.

Look no further than your last red-eye flight or talk to the parents of newborn children, and you'll see what a lack of sleep does to our capacity to deal with things.

That "short emotional fuse," as a 2015 Israeli study calls it, may be due to increased amygdala activity. Yes, the same amygdala that is part of the limbic system we're trying to override with everything we've discussed in this book. [61] Sleep, or the lack thereof, has a measurable and immediate negative impact on our states, leads to decreased performance, and deleteriously alters course when chronically dysfunctional. [62, 63, 64]

Consider this personal experiment conducted by neuroscientist Dr. Sarah McKay, discussed in her TEDx talk "Indulge Your Neurobiology."

Dr. McKay intentionally altered her sleep with light exposure and measured how the lack of sleep impacted her brain. She experienced significant decreased motor control (she dropped acid in the lab more than once) and "felt out of control emotionally." [65]

Our anecdotal experiences, and that of Dr. McKay, are supported by research on sleep and its impact on HRV, happiness, health, and decision-making.

Better sleep increases HRV

Subjects with sleep apnea were treated with a CPAP machine; the resulting improvements in sleep led to an acute increase in HRV after a single night. [66]

A 1997 study of college students concluded that sleep quality is more important than quantity. Both of these make sense. Sleep is a restorative, mostly parasympathetic activity. As we've seen, more time in these PNS states increases HRV. [67]

Another study, this one from 2013, found that low-quality sleep makes the amygdala up to 60% more reactive (not a good thing – remember that the amygdala is the emotional driver of the limbic system). We want the prefrontal cortex and our higher levels of consciousness running the show. [68]

Sleep is when our brains remove the metabolic waste from the previous day. This waste includes a compound called adenosine, a natural byproduct of healthy functioning. The problems arise when that builds up because it has not been cleared during our sleep. We feel less alert, less mentally sharp, and "foggier." [69]

Sleep, health and happiness

In a Harvard Health Review, inadequate levels of sleep were linked to higher levels of stress (lower HRV) and higher incidences of depression. [70]

Using both electroencephalography (EEG) to

measure brain waves and functional MRI to view neural activity, the Israeli researchers mentioned at the top of this section found that decreased sleep lowers the threshold for emotional activation and leads to increased anxiety and reduced emotional resiliency. [71]

Sleep disordered breathing, which has previously been linked to reductions in sleep quality, has been so strongly linked to dysfunctions in glucose intolerance and insulin resistance that it may lead to type 2 diabetes, according to a 2004 study. [72]

Sleep and decision-making

Sleep experiments conducted by psychologist David Dinges, PhD, revealed that "people who get fewer than eight hours of sleep per night show pronounced cognitive and physiological deficits, including memory impairments, a reduced ability to make decisions and dramatic lapses in attention." His research also showed that these negative effects compound over time. [73]

Sleep improves learning and memory

Sleeping has been shown to help us process and retain information. In fact, sleeping or taking a nap after learning new material helps improves memory consolidation and performance after sleep. [74, 75]

This exciting study published in 2016 found that HF (high frequency) HRV during sleep helped consolidate memories. Researchers "hypothesize that central nervous system processes that favor peripheral vagal activity during REM sleep may lead to increases in plasticity that promote associative processing." [76]

As part of our "everything is everything" theory,

sleep increases HRV, emotional resiliency, and neural plasticity. Neural plasticity, as we'll explore, helps us rewire the desirable habits and traits we want (need) for long-term embodiment of the elevated states of consciousness in which we seek to live.

The image below is a great diagram of the interrelated neurovisceral, vagal nerve integrations we've been discussing through this book.

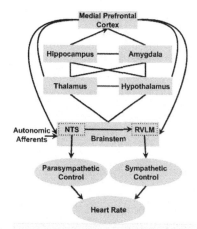

Fig. 1.

Central nervous system memory-related areas and autonomic control of the heart. ANS fibers from major organs in the periphery, including the heart, innervate the brainstem at the nucleus of the solitary tract (NTS). The brainstem connects with the thalamus, hypothalamus, and other memory-related areas, including the hippocampus, amygdala, and PFC. Descending projections from the PFC to the hypothalamus and brainstem create a feedback loop facilitating the modulation of peripheral activity by the central nervous system, including HRV. Lines denote bidirectional connections, and arrows denote monodirectional projections. Note that, for clarity, we only show a partial representation of the central autonomic network in the current figure. RVLM, rostral ventrolateral medulla. Adapted from ref. 12.

Tips to enhance sleep

Sleep is crucial, but how can you ensure consistent, high-quality sleep?

It starts with a focus on what we call "sleep hygiene." Also known as "bedtime routine," the effectiveness of this ritual lies in the power of conditioning. Like Pavlov's famous research with salivating dogs, we become conditioned to prepare for sleep if we do the same thing each night before going to bed.

Every expert and website on the Internet has their own take on what this bedtime routine should look like. I say sample many, keep the elements that work for you, and discard the rest.

Remember it's not about how many things you can add to a routine; it's about getting better, more consistent sleep. Don't lose sight of your main goal.

This routine should help you wind down, shift your body from sympathetic to parasympathetic state (also called "rest and digest"), and, over time, condition your brain and body to begin this shift at the same time every day.

Some common parasympathetic-inducing activities include taking a warm bath, using essential oils like lavender or eucalyptus that are known to promote calm and relaxation, doing breathing exercises, reading a book (avoiding screens and maybe avoiding non-fiction books that "stimulate" your mind), meditating and doing a gratitude journaling session. [77]

Perhaps the two most important things to do (or avoid) to promote better sleep are avoiding blue light after dark and avoiding caffeine after a certain time. That caffeine cut-off time will vary from one individual to the next, but I typically recommend 1-2pm as the latest.

After dark, try to limit blue light exposure by wearing blue-blocker glasses, using programs like F.lux or Iris on your computer and phone, and limit screen time as much as possible. This will be a lifestyle shift for many, but remember: "Your life is perfectly designed for the results you're currently getting." If you want better sleep, be prepared to make some shifts.

The ultimate sleep environment is one area where I disagree with many. My recommendations are based on conversations with sleep researchers, naturopathic doctors, and ancestral health experts.

The main idea they all agree on is that our ancestors slept in cycles. The old and young normally went to bed first, followed by the adolescents and women. Some men then went to bed, but others stayed up as sentries to guard the tribe. The elderly usually woke before the others. Of course, there were (and are) individual variations, but the point is that there was usually a fire burning and someone stirring about.

So the common recommendations that you must sleep in absolute darkness and absolute silence are not serious concerns for me. I'd rather you focus on reducing blue light after dark and limiting caffeine to get high-quality, restorative sleep.

There is research to support sleeping in a colder than average room. Studies put the optimal range between 62 and 68 degrees Fahrenheit, with 66 being the ideal number. [78]

I also recommend shutting off your Wi-Fi at night. You're not using it, and there is evidence that constant exposure to it could be harmful to our health. An even better option would be to ditch your Wi-Fi and use hard-wired ethernet cables, but I realize most will not do that. So use one of those holiday light timers and set your Wi-Fi to turn off from 10pm to 6am every day (or whenever you sleep). Eight hours a night will save you nearly 120 days of exposure over the course of a year.

If you're going to put all of this effort into improving sleep quality, you might as well track it. And the best way to track it is HRV. Despite everything written in this book, I have yet to find an HRV measuring device that I'm happy with. They all have flaws and frustrations, and the only reason I put up with them is the valuable data; it's like waking up and trying to shuck a raw oyster with a butter knife.

I've used a finger sensor that was hard-wired to my phone and a Bluetooth heart rate monitor, but both frustrated me to the point where I stopped using them daily. For this reason, I don't track HRV every day. I will check it using one of these two methods once or twice a week, if the technology cooperates.

The Oura ring is a friction-free method. It's worth noting that the measurement provided (at this point) is not a true HRV. They describe their "readiness" score as a composite value based on 24-hour rolling HRV along with other data it collects.

You can also track your sleep with something like Sleep Cycle or Beddit, but again, I'm not a fan of Wi-Fi exposure during sleep. To me, not knowing is better than the alternative of increased exposure.

It's worth noting that both Brain.fm and neurofeedback can improve sleep quality. Brain.fm boasts a 2 to 4X increase in brain function, with increased slow-wave sleep when using their sleep music. You'll need special sleep headphones that you can purchase on Amazon, and be sure to download the sleep track so you can listen with your phone on airplane mode and the Wi-Fi /Bluetooth antennas turned off.

Throughout the book so far, we've mentioned HRV several times without defining exactly what it is or why it's so important. We're going to dive into that now, and you'll realize how inextricably linked all of our body's systems truly are.

CHAPTER 4:
Polyvagal Theory and Heart Rate Variability

Get More Play

When you're young, you do it because you love it.
Never grow up, my friend.

As much as that sounds like a line from *Peter Pan*, it's not. It's actually from a Nike soccer commercial that aired during their "Joga Bonito" (play pretty) campaign featuring Brazilian superstar Ronaldinho circa 2008-2010.

That video resonated with me then and has never been very far from my mind in the years since. The goal of the narrator in the commercial was to "remind us how it should be," as it spliced a video of the soccer superstar in his present form with a video of him as a child phenom – always smiling and playing for fun, not the stress of winning and pleasing the critics as many athletes do as their careers advance.

There is something infectious about a person at play and the genuine smile that accompanies it. One of the most blissful, light-hearted states that I have enjoyed in recent memory occurred last June at my in-laws.

As we took my wife's father to get his weekly haircut and shave at my sister-in-law's, we walked past a basketball goal lowered to eight feet, and a basketball was just sitting there, almost teasingly. It was begging me to pick up the ball and dunk it.

So I did.

Then I did it again.

And again.

Then I started adding reverse dunks, self-alley-oops, and tossing the ball off the backboard.

What seemed like two minutes went by and, all of a sudden, they were done inside, and it was time to go. I hadn't even noticed the sweat, the heavy breathing, or the passage of time until it was over. But I sure as shit noticed the smile on my face and how much fun I was having.

I was transformed.

I'll admit it, I'm a hard-charging, type A, "discipline equals freedom" type of person. While that serves me well more often than not, it has the obvious downside of not playing enough. Getting enough spontaneous play in my life is something I am (and will continue to) work on.

Having that light-bulb, a-ha moment on the eight-foot basketball goal at my in-laws was a powerful shift for me. That moment stands out because it was the first a-ha moment in regards to play and having fun. There have been many such moments since then, but only because I have been aware of the need to play and actively seek those opportunities.

Perhaps more crucial is the fact that I now *allow* myself to let loose, play, and lose myself in those moments. I no longer feel guilty about taking a half day off from work on a weekday to go surfing.

If you're not able to let go of the feelings of guilt, you're not being fully present. And you're also not allowing the vagus nerve to feel safe – both of which defeat the point.

Let's talk about why play, safety, and something called the vagus nerve are crucial for our well-being and performance.

The Polyvagal Theory

Understanding the polyvagal theory was truly my "everything is everything" moment. First posited by Stephen Porges, this theory is yet another scientific discovery showing that our physiology dictates our behavior.

It's been difficult to delay talking about the polyvagal theory and heart rate variability (HRV) until now as it is truly a useful model to explain how we function and operate. But I wanted to cover the previous material first for two reasons.

Number one, the theories and works presented so far pre-date the polyvagal theory and provide a foundation of neurological understanding of the layers and the importance of Porges's work. You'll also be able to see how these systems overlap and impact each other, especially the sympathetic fight or flight response.

Second, I held the polyvagal theory until now because I think it is the game changer. As I said, it's my "everything is everything" moment. I think it could be the light bulb, "a-ha" moment that will help many connect the dots. Its implications and applications are vast, and I want to be able to run with them, so let's jump in.

Simply put, Porges's theory is the neurovisceral connection that writers, philosophers, neuroscientists, and psychologists have been observing for centuries.

Our bodies and brains influence each other via a two-way street, and that two-way street is the vagus nerve.

Stephen Porges didn't discover this nerve – far from it. Remember Darwin's excerpt on the two-way communication of the vagus nerve, written in 1872:

"When the heart is affected it reacts on the brain; and the state of the brain again reacts through the pneumo-gastric [vagus] nerve on the heart; so that under any excitement there will be much mutual action and reaction between these, the two most important organs of the body." [79]

While he didn't discover these unique functions of the vagus nerve, Porges was the first to decode the polyvagal paradox. In his words, Porges describes the significance this way: "It's neurobiological science meets behavioral science – they are related! They influence and impact each other. They're inseparable." [80]

Let's wander through this. The vagus nerve, or cranial nerve X, is a meandering nerve so named because it wanders like a vagabond as it carries parasympathetic (rest and digest) information to the visceral organs of the digestive, respiratory, and urinary systems. [81]

The vagus nerve is a mixed nerve, meaning it conducts both incoming (afferent) sensory information and outgoing (efferent) muscular commands in the same nerve bundle, hence the "two-way street" analogy.

It innervates and connects the brain with the heart, lungs, liver, stomach, and intestines, to name a few, and it plays a major role in heart rate and digestion. What makes the vagus nerve so special is that it's not just a one-way street.

As Darwin and so many others have noted, the information traveling in both directions along the vagus nerve impacts and directs physiological functions at both ends. The brain talks to the visceral organs, and those organs also send information to the brain.

The Vagal Paradox

Stephen Porges first posited his polyvagal theory in 1992, when he was faced with seemingly paradoxical data on the vagus nerve.

On the one hand, Porges had evidence that the vagus nerve was a nerve complex that, when activated, triggered immobilization as a defense mechanism.

On the other hand, he also had compelling data that the vagus nerve was contributing to cardiorespiratory index – something he called vagal tone (which we call HRV).

In his work with babies, Porges was using this cardiorespiratory index as an indicator of overall health, and he found that pre-term babies had a lower index than full-term babies, indicating that vagal tone was linked to cardiorespiration.

Porges was faced with a question: which proposed function was correct? It turns out both were correct, and the Poly (Multiple) Vagal theory was born.

Thanks to Porges, we now know that vagus nerve is one nerve with two separate functions. The older, more primitive pathway became known as the dorsal vagal complex. This complex is unmyelinated, meaning it lacks the fatty myelin sheath that encircles and insulates most nerve cells. Myelination improves the speed of electrical impulses, making unmyelinated nerve complexes slower than their myelinated counterparts.

Almost all vertebrates have this dorsal vagal complex, and it is sometimes known as the "vegetative vagal complex" as it is the most primitive vagal function and is linked to immobilization when faced with a threat; this is why many lower vertebrates, even some mammals, will freeze or immobilize when this defense mechanism is activated.

The latter vagal conduit became known as the ventral vagal complex. This more recently evolved complex is seen only in mammals and contains faster, myelinated nerves that can influence cardiorespiratory function and can inhibit the limbic system, depending on the situation (and as we're discussing in this book, level of cognitive control/emotional resiliency/training).

As you might imagine, the secret to optimizing this system is achieving balance, even tipping the scales in the favor of shutting down the evolutionary, survival-focused defense mechanisms that cloud our higher-level thoughts and decision-making.

Porges outlines the normal functioning of the vagus in a triage-like phylogenetic hierarchy, asserting that we use the newest conduit first, and if that doesn't work, we go down the line in succession from newest pathway to oldest seeking a solution.

According to Porges, "The hierarchy emphasizes that the newer "circuits" inhibit the older ones. We use the newest circuit to promote calm states, to self-soothe and engage. When this doesn't work, we use the sympathetic-adrenal system to mobilize for flight and flight behaviors. And when that doesn't work, we use a very old vagal system, the freeze or shutdown system."

What does this look like in real life?

In this sequence, when faced with a threat that activates our defenses, we would first rely on the ventral vagus cortex.

This is the "normal," evolved, socially engaging pathway that will use facial features, expressions or even verbal communication to determine threat levels and possible solutions.

For example, after bumping into a large stranger in a crowded bar, we may try to immediately de-escalate the situation with open-palmed gestures, an "excuse me" and an offer to replace the drink we spilled.

If that doesn't work, we move down the hierarchy to the dorsal vagal cortex and sympathetic response. Faced with a fight or flight scenario, we literally have to choose to fight or leave the scene. Just as the ventral vagus cortex is associated with normal functioning of the digestive system, this state of fight or flight is associated with constipation. Our body has diverted resources away from the brain and digestion and towards the extremities for action.

In extreme cases, we may become paralyzed with fear, revert to extremely primitive behaviors, and curl up in the fetal position on the floor, immobilized. This might also look like playing dead if we encounter a grizzly bear during a hike in the woods. The immobilization defense response is associated with uncontrolled defecation.

This bar scene is an extreme example, and certainly does not reflect the way we experience the polyvagal theory on a daily basis (hopefully) but it clearly shows its triage-like hierarchy. While the constipation and uncontrolled defecation seem like extravagant details, their relevance will be revealed as we discuss our more realistic, daily encounters with the polyvagal theory through low-level, chronic exposure to such defense responses in our lives.

As with everything we've discussed to this point, it's about normal function versus dysfunction. Since we're primitive beings wired for survival, living in a modern world, our biology can get us into trouble.

Dorsal Ventral Complex Run Amok

Under normal circumstances, the dorsal ventral complex of the vagus nerve governs subconscious control of the digestive processes of our visceral organs.

This is the same pathway that, when activated for defense, initiates our sympathetic fight or flight response.

We've already talked about the problems associated with chronic low-level activation of the limbic system in the last section (elevated heart rate, low vagal tone/HRV, disrupted sleep or difficulty sleeping, constantly anxious, irritable, stressed out, and a general lack of ability to deal with the things life throws at us, to name a few).

What's interesting to me is the literal constipation that fight or flight brings and the ensuing emotional, spiritual, and "life" blockage or constipation that corresponds to living in a constant state of panic (that's what fight or flight is).

Our inability to shut down this defense response leads to literal and metaphorical blockages.

Beyond the sympathetic fight or flight lies the even more primitive pathway of immobilization. Immobilization as a defense can be seen in anyone who shuts down from trauma.

Porges notes that in non-life-threatening situations, we often see this as dissociation. [82]

How many people have you seen sleep-walking through life because they're "scared shitless" (no coincidence this phrase is used – recall that immobilization is linked to uncontrolled defecation) to face the perceived threats in their lives?

I'm not talking about post-traumatic stress disorder; that is a sensitive topic and one for another book, but on a subclinical level, there are millions of people feeling stuck (or frozen) in their lives because their vagus nerve is relying on their most primitive defense mechanisms to deal with issues they're not facing.

People staying in jobs they hate or unhealthy relationships are immobilized by their primitive lizard brain.

Speaking of living in dissociation, Porges also notes that much of modern medicinal intervention shuts down feedback loops for us to feel ourselves, creating more of this dissociation and immobilization.

Needless to say, the polyvagal theory has massive potential applications in healing and wellness. But for the scope of this book, we're focused on optimization for healthy people. (This is as good a time as any to remind you that this is not a medical advice guidebook).

To anyone living in either of these states, immobilized by fear or constant compulsive, self-sabotaging, dopamine-seeking, limbic system-driven behaviors, I say this: I've been there. I know how frustrating it can be to be stuck in a seemingly never-ending vicious cycle of self-sabotaging behaviors and negative feelings, all of which lead to more regrettable decisions made for short-term relief.

The good news is this: you can right the ship, and you can train your mind to be equally (or even more) powerful on the positive side of this cycle. You're learning how the mind works now, and I'll lay out the exact blueprint for you to successfully F*ck Your Feelings, Move the Chains, and direct your decisions the way you want – all day, every day.

Heart Rate Variability

We can't talk about polyvagal theory, vagal tone, and emotional resiliency without discussing HRV.

Discovered by Porges in the 1960s, HRV is the variability in time between heart beats (not to be confused with the actual time between the beats).

In other words, the time between our heart beats is not always the same amount of time. That variance in time is HRV, and it's synonymous with the term "vagal tone," making HRV an indicator of what's going on with our vagus nerve and how well we're balancing our time in the sympathetic and parasympathetic states.

When Porges first published his findings on HRV, he recalls being ridiculed by other scientists who were so naive about this neurological function that they dismissed his findings and said his readings were because he was "a bad scientist."

From the beginning, Porges used the terms vagal tone and HRV interchangeably, so I will too. Over the last few years, HRV has exploded in popularity as a biomarker and quantifiable metric of overall recovery, health, and readiness. The scientific and medical communities have used it to quantify mortality rates for things like cardiac and kidney diseases, with lower HRV being established as a significant independent risk factor for higher mortality rates in diseased and healthy populations. [83]

Many athletes use it to quantify recovery and look at it as real-time feedback to plan the day's training for best long-term results.

What's interesting is that nearly every "recovery activity" used by athletes to increase HRV – which we now know increases vagal tone and promotes emotional resiliency – is an activity that quiets limbic system functions, shuts down defense responses, and moves us from sympathetic states to parasympathetic states.

One study noted that HRV fluctuates with respiration – increasing with inspiration and decreasing with expiration – and is primarily mediated by parasympathetic activity. [84]

In other words, my "everything is everything" theory is that monitoring HRV and working to increase it, both short and long-term, could lead to increased emotional resiliency and possibly a decreased incidence (and certainly a decreased magnitude) of negative emotions, feelings and thoughts.

The research supports this, with multiple studies finding a link between increased vagal tone (HRV) and demonstrably increased emotional resiliency.

Noted psychophysiologist and behavioral researcher Julian Thayer has been investigating the link between HRV and behavior longer than most. In 2002 and 2009 he published research, along with others, that found "higher resting HRV is associated with the effective functioning of prefrontal-subcortical inhibitory circuits that support flexible and adaptive responses to environmental demands." [85, 86]

In 2012, Thayer performed a meta-analysis of neuro-imaging studies on the relationship between HRV and regional cerebral blood flow.

If you had to guess which areas of the brain showed up on their brain imaging, what would you say? If you guessed the amygdala and ventromedial prefrontal cortex, you'd be correct.

Here's what Thayer and his team wrote about the results of that 2012 meta-analysis:

"We further propose that the default response to uncertainty is the threat response and may be related to the well-known negativity bias. Heart rate variability may provide an index of how strongly 'top-down' appraisals, mediated by cortical-subcortical pathways, shape brainstem activity and autonomic responses in the body. If the default response to uncertainty is the threat response, as we propose here, contextual information represented in 'appraisal' systems may be necessary to overcome this bias during daily life. Thus, HRV may serve as a proxy for 'vertical integration' of the brain mechanisms that guide flexible control over behavior with peripheral physiology, and as such provides an important window into understanding stress and health." (87)

If you're not a geek like me who enjoys reading abstracts and scientific journals, here's what that means: the human default response to uncertainty is the threat response, aka the limbic system, as we mentioned in previous sections of this book.

This may be related to the well-known "negativity bias," or as we discussed, the human survival skill of always looking for what's wrong.

HRV is a quantifiable, real-time metric that can tell us how capable we are of "top-down" control, i.e., the conscious mind attenuating defense response mechanisms and helping us act rationally rather than emotionally.

Building on Thayer's earlier research, we know that increased HRV leads increased capacity to deal with uncertainty, potential threats, or daily stressors.

In regard to F*cking our feelings, increased emotional resiliency translates to a significantly increased capacity to deal with shit. This is our bandwidth capacity quantified, or our threshold at which the prefrontal cortex loses control and allows the limbic system to run amok.

So what do we do with this amazing insight?

Vagal Tone

The term vagal tone can be used interchangeably with HRV. So high vagal tone, or high HRV, means you have a greater bandwidth or greater capacity in that moment or on that day to deal with things. This is how it leads to increased emotional resiliency.

Remember that HRV is (at the risk of oversimplification) a balance of how much time we spend in the parasympathetic versus sympathetic state. Most people today spend way more time in the sympathetic state than the parasympathetic, which is why our recovery, health, and emotional resiliency are in constant states of overtaxed.

How can we spend more time in the parasympathetic state? With any parasympathetic activity, like sleeping, taking a nap, breath work, cold exposure, yoga, meditation, float tanks, gratitude journaling – really anything where you slow down, destress, disconnect from that fast-paced, frantic life that we live now.

<u>Implications and applications</u>

I've been a big fan of increased time in the parasympathetic state as a means to enhance recovery for years, but the idea is that it can significantly impact awareness and consciousness, and it is potentially game(life)-changing.

The mathematical transitive property of equality states that "if $A = B$ and $B = C$, then $A = C$." Through this law, everything we can list as an activity that increases HRV (A) can also be employed as an activity to increase vagal tone (B), or emotional resiliency (C).

In other words, more time in the parasympathetic

state (which by definition takes us out of sympathetic fight or flight, aka limbic system activation) can increase emotional resiliency and help us gain conscious control of our emotions, feelings, and decisions.

As was so elegantly stated by Dr. Raleigh Duncan when I interviewed him for my podcast, "The body will not devote resources to recovery if it thinks it's being chased by a saber-toothed tiger." While we may not encounter man-eating carnivores on our daily outings anymore, the modern world has us living in chronic, low-to-mid-level sympathetic activation.

There are numerous tools at our disposal to switch off this threat/defense response and get our HRV trending in the desired direction. These parasympathetic activities: a) directly enhance HRV, b) provide opportunity to train our prefrontal cortex to become the default pathway, c) down-regulates overactive defense response. It's all a self-perpetuating system – breathe, meditate, increase awareness, control your mind, and it becomes easier/stronger the more we do it.

Making behavioral changes is always challenging as we're literally rewiring our brains to create new habits. Remember our previous discussion of carving a new road and turning it from dirt road to highway through thousands of repetitions.

Be patient with yourself. You won't master this on the first or second attempt. This is a lifetime pursuit. There is a reason monks live in solitude to master this very skill.

Give yourself time, and if you slip up and use the old wiring system, don't beat yourself up. Acknowledge it (awareness), accept it (you can't change the past), forgive yourself (so you don't become your own threat-inducing enemy), and move on. Focus on what you can control: your attitude and efforts in future situations.

Getting started and being consistent will be your two keys to success. Look at any successful person and you'll see they did exactly this. Many of them started from humble beginnings with seemingly insurmountable odds (Dr. Dre surviving the streets of Compton, Springsteen's Mom borrowed money for his first guitar...), and they displayed remarkable consistency that would shame the rest of us if we took an honest look at our work ethic.

Simply put, they refused to stop. If we apply a fraction of this commitment to our pursuit of self-improvement, we'll experience exponentially more rewarding lives.

So get started, fight through the challenge, be consistent, and become a resilient, positive, badass motherf*cker.

Living with Elevated HRV

Recall our states and traits, or climate vs. weather discussion from the beginning of this book. We can apply this concept to HRV, and therefore our decision-making, time management, and all that follows.

It's about traits. It's about the norm, not the exceptions.

What do you do *most* of the time?

We want high vagal tone to be a trait, like that of Florida's climate.

Sure, life will throw storms at us. California deals

with earthquakes and wildfires, Florida deals with the occasional hurricane — we too will deal with low HRV days, down days, etc.

Developing the traits outlined here will help you reduce both the incidence and severity of those days. Then, when those days *do* occur, we can lean on the physiological "hacks" to positively alter our states.

Living with increased HRV is the same as increased vagal tone, which we know leads to increased emotional resiliency. Think of HRV as a real-time look at your bandwidth status. The lower your HRV, the less bandwidth you have to deal with shit, and the more likely you are to stray from your chosen path.

Living with increased awareness and in heightened states of consciousness both relies on higher vagal tone and influences it. Like the vagus nerve itself, it's is a two-way street.

The Vagus Nerve and Chakras

If you haven't noticed by now, I'm an incredibly curious person. Knowing that the vagus nerve touched many of our visceral organs, I wanted to test my "everything is everything" theory by looking at how the vagus nerve aligned with our chakras.

Generally associated with meditation, yoga, and traditional Ayurveda, chakras are defined as the centers of spiritual power in the human body.

The word comes from the Sanskrit word "cakra" for wheel or circle, and chakras are generally thought of, and depicted as, vortices of energy.

What's even more interesting is that Eastern medicine and acupuncture recognize many of the same regions as crucial nerve complexes for energy and healing.

Both systems are focused on the movement and/or blockage of energy within our bodies. Both systems envision two snakes wrapping around the spine, where each intersection is a chakra or meridian.

This same two-snake imagery is seen in the American Medical Association's logo:

There was too much coincidence/overlap to not include this information in the book, but some of the correlations are anecdotal at best. That's why it's an anecdotal insert rather than a chapter in itself.

Below is a table I put together showing the seven chakras and their corresponding Eastern (acupuncture) and Western (vagus nerve plexus) counterparts.

The seven chakras, their colors, musical notes and acupuncture relatives are shown in this table.

Chakra	Vagus Nerve Link	Chord	Color	Function	Acupuncture Point
Root	Celiac Branches*	C	Red	Feelings/Survival	Hui Yin
Sacral	Pelvic Plexus	D	Orange	Acceptance/Abundance/Pleasure	CV3-8
Solar Plexus	Celiac/Hypogastric Plexus	E	Yellow	Self-worth/esteem/confidence	Zong Wan
Heart	Cardiac Plexus	F	Green	Love/Joy/Peace	Dan Zhong
Throat	Laryngeal Nerve	G	Blue	Communication/Truth	Tian Tu
Brow/3rd Eye	Ciliary Ganglion	A	Indigo	Intuition/Wisdom/Big picture	Yin Tang**
Crown	Sphenopalatine Ganglion	B	Clear	Connection/Bliss	Bai Hui

- * Colon innervation by the vagus nerve remains somewhat controversial, but a 1993 paper "indicates that all regions of the colon, except the rectum, are innervated by the celiac and accessory celiac branches of the vagus nerve." [88]
- **It's worth noting that the sixth chakra, the brow, or the third eye, is noted as an "extraordinary point" in acupuncture. It's associated with the pineal gland, a brain structure that resembles the all-seeing Eye of Horus from Egyptian inscriptions. This structure is quite literally a third eye in lower organisms, and Western science is exploring its potential as the seat of human spirituality. [89]

Until recently, I was unaware that the chakras had associated musical notes. I find this fascinating, since the connection between mood and music has long been established –something we'll explore it in greater detail later in this book.

But if we examine musical notes for what they are(pure sound vibrations), it's not a stretch to think those vibrations can influence the vibrations of our cells; after all, the way we "hear" is through the interpretation of vibration in the bones of our ears.

And as mammals, our friend Steven Porges reminds us, we have hearing adaptations associated with our highest level of polyvagal theory (communication) that allow for the interpretation of spoken communications that lowers organisms do not possess. (For example, snakes only hear pounding vibrations from footsteps and must interpret threat/no threat based on that alone. They cannot hear our spoken communications the way we do).

And at the atomic level, all we are is atoms, energy, and vibration, and these frequencies could provide instant and immediate syncopation of vibrations, helping to "tune humans" as one might tune an instrument.

This is similar to the science behind binaural beats, entraining the brain to produce desired brain wave frequencies based on musical notes and frequencies of vibration – something we'll discuss soon in the "Brain Waves" chapter.

Aligning chakras and paying attention to their correlation to vagal innervation and acupuncture meridians as a way to potentially increase vagal tone, heighten awareness/consciousness, and drive increased emotional resiliency is something I'm experimenting with and seeing tremendous benefit.

So far, however, the only conclusive science I can find to support the increased HRV comes from meditation/breathing and increased time in the parasympathetic state. There are reviews of acupuncture and HRV, but the results are mixed, and more studies are underway. Maybe soon we'll have studies that support the "woo-woo" ayurvedic chakra alignment and acupressure meridians as evidence-based methods to increase HRV. For now, I'll list meditation as a parasympathetic activity and stick to the research-backed claims. [90]

Reflection Questions

- Are you tracking HRV? How can you use this data to increase your awareness and strengthen emotional resiliency?
- In what circumstances do you feel your emotional bandwidth disappearing? How can you protect it? How can you increase it?
- Does your community celebrate you? Or tolerate you?
- Do you feel safe taking moonshots?
- How often do you drop everything and play? How balanced does your life feel? How can you make play a bigger part of your life?
- Is your tribe or community making you play small? Or are they encouraging you to excel, grow, and evolve?

Choose your community wisely

"We are the sum of the five people with whom we spend the most time."

This age-old adage has survived for so long for one reason: it's f*cking true.

We've all experienced the power of this truth in some form or fashion.

Maybe we hung out with the wrong crowd as

teens, or saw a friend go down that road. Hopefully, you can look back on times in your life and realize how fortunate you were at certain intervals of life to be surrounded by greatness and how it helped propel you forward as a human.

If you haven't experienced this (yet), know that it is life-changing, and it's never too late.

In fact, the endeavor to surround ourselves with like-minded, positive, motivated people is a never-ending audit to ensure that our group grows and evolves with us. An equally important but often overlooked component of this is that these communities should be vigilant about avoiding or culling those people who bring down the group as a whole.

Every successful athlete and entrepreneur echoes the importance of this rule. Somehow, we innately understand this concept, yet so few of us embody it in our daily practice to the extent that we might wish.

Why we ignore this is another issue in itself, but it likely has to do with comfort.

We're comfortable in our current situations – not motivated enough to make a change (more on this later). We're also wired to exercise caution when it comes to alienating ourselves. Literally, our biology is hardwired to enforce group norms and keep us safe inside those constructs.

To call out members of our social circle, or even scarier, to strike out on our own, was a death sentence in ancestral times, when we required the tribe to survive. Despite our ability to survive without others in the modern world, the thought of alienating ourselves from our social group seems just as scary, especially if we have not identified the new group with which we will associate.

To this, I advise you: find the people you want to be around and go there. Befriend them and make them your new tribe. Neuroscientists and behavioral scientists agree.

Northwestern University neuroscientist Moran Cerf has been studying decision-making for over a decade, and he asserts that the single greatest influence on our happiness and success in life is the company we keep.

Choose your community wisely, for the values of your community become your values.

Nerf asserts that if we want to maximize happiness and minimize stress, we should surround ourselves with people who embody the qualities we value. Over time, we'll naturally adapt these desirable behaviors and traits, and at the same time reduce the demand on our brain for making decisions. [91]

This last part scares me, and I included it as a caution. Recall our discussion from the previous section about awareness and conscious choice. Do not become a mindless sheep following the current without doing your own research and your own thinking, no matter how esteemed your circle of friends may be. Never lose yourself and your unique character, and most importantly, your unique thoughts and beliefs. Stay centered in your core values and expand in alignment with those values.

Getting back to Cerf's research, the way we adapt the behaviors and traits of our social circle is attributed to something called "mirror neurons."

Mirror neurons are located in the prefrontal cortex, and their full purpose is still being investigated – as is the case with most things involving the brain and behavior. Currently, scientists speculate that they're involved in learning new skills through imitation. [92]

What we know at this point is that there are many different types of mirror neurons; the research is not complete, and many studies are refuted as they've been conducted on monkeys.

In a 2005 study, UCLA neuroscientist Marco Iacoboni had human subjects watch someone pick up a cup of tea from a table and asked them to determine whether the intent was to clear the table or actually pick up the tea and drink from it. His findings – that subjects could tell a difference in the intentions – led him to the conclusion that mirror neurons played a role in empathy and understanding the actions of others. [93]

Both Cerf and Iacoboni, along with many other behavioral neuroscientists, are simply quantifying and understanding the mechanisms behind what we already know – that we become the people with whom we surround ourselves.

Consider the following scenarios.

In the first scenario, your four closest friends all watch *Game of Thrones* on HBO, or maybe they each take fantasy football very seriously, flying across the country for draft-day parties, and spend hours each week strategizing over how to score the most points in an effort to win the league.

Chances are, if you become part of this tribe and don't already watch *Game of Thrones*, you're going to check it out, since everyone around you is constantly talking about it. Maybe it wasn't *Game of Thrones*, but I'm willing to bet you've started watching a new show, listened to a new podcast or checked out some musician based on the fact that several of the people in your tribe mentioned them. The same applies to fantasy football or anything that continuously dominates the time and conversations of those around you.

They value those activities, and either you will adopt those values or migrate to a new circle of friends/influence.

Now consider an alternative scenario in which none of your four closest friends even own a TV. All four have side businesses on top of their primary source of income, where they make extra money from their passions. All four are fanatical about health – eating at healthy restaurants and cooking healthy meals at home with foods they picked up at the local farmer's market – and all four regularly work out.

What behaviors do you think will seep into your life? Are you already thinking about the passion you would pursue and try to turn into a side business? If so, you're aware, and it's that awareness that creates choice.

If you begin to hang out with people who spend their time at a bar and drink copious amounts of alcohol, it's likely that you will begin to do the same. If you start hanging around billionaires, your view of the world will change. Your mind will be exposed to new possibilities, and your belief systems will expand beyond anything you've ever imagined.

As we've established in the above scenarios, your interests, choices, values, and activities (aka, how you spend your life) will vary greatly based on the people you keep around you.

You are the sum of the five people with whom you spend the most time.

Choose your fiends wisely.

If you can't physically do this, you can still do it mentally and emotionally.

Tom Bilyeu, co-founder of Quest Nutrition and host of Impact Theory, told me that he used books to upgrade his social circle. Since he could not physically be with certain people, he isolated himself from the negative people in his geographical location and immersed his mind in the mental world of those whom he admired and valued most.

I've created the Move the Chains Academy for this exact purpose. The Academy is where hundreds of like-minded people form an amazing community of support an accountability to implement the strategies in this book, in an effort to elevate awareness, consciousness, Move The Chains, and level up their lives. For an instant upgrade in your social circle, join the #MTC Academy at ryanmunsey.com/academy.

In early 2016, Charles Duhigg, the best-selling author of *The Power of Habit* and *Smarter. Faster. Better* published the article *"What Google Learned From It's Quest To Build The Perfect Team"* in the New York Times that changed the way many viewed collaboration and success. [94]

Google had just finished a two-year study evaluating 180 teams, looking for the secret recipe to create high-performing, more successful teams.

As Duhigg explains, the researchers were stumped. They struggled to identify why some groups excelled but others fell behind. That is, until they dug deeper into the research and started looking at "group norms."

As they struggled to figure out what made a team successful, Rozovsky and her colleagues kept coming across research by psychologists and sociologists that focused on what are known as "group norms" – the traditions, behavioral standards, and unwritten rules that govern how teams function when they gather. Norms can be unspoken or openly acknowledged, but their influence is often profound.

Google finally published their findings, noting that the highest-performing teams shared the following five qualities: dependability, meaning, structure/clarity, impact, and the one that stood out most – psychological safety.

That's right, members needed to feel safe in their ability to voice opinions, share feedback, and push the boundaries of creativity. This, according to Google, was the single biggest factor in determining whether or not a team would excel or fall behind.

Google isn't alone. In a recent conversation behind the stage of a large conference, Peter Diamandis reiterated to me the importance of "needing to feel safe in order to take moonshots." On stage, he posed the question: "Are you celebrated in your community or merely tolerated?" His point being, that folks who are merely tolerated feel less safe and are therefore less likely to operate at their fullest potential. They'll play small in order to avoid rocking the boat.

Do you feel comfortable sharing your wildest dreams with your community? Do you have a tribe? Do they tolerate or celebrate you and your ideas?

The notion that feelings of safety could be linked to success was an interesting concept for me and many others who read this viral story.

I kept coming back to this safety thing, how it impacted our vagal tone, and how that, in turn, inhibits or encourages our ideas to explore new frontiers.

The dichotomy of this phenomenon is fascinating. We must feel safe in certain areas of our lives in order to play big, take risks, and expand our comfort zones in other areas.

After all, pushing our boundaries is what this book is all about. We all want the proverbial "more" – not in the material sense of the word, but in the life fulfillment and satisfaction realm.

We want to be, do, and achieve more. We know we're capable of five, ten, twenty times more than we're doing right now, and we're seeking that crucial piece of information that will help us tap into our authentic purpose and coax out our inner greatness.

This is directly in line with the question posed as I set out to write this book.

Why do some people achieve massive success, while so many others struggle, toil, stagnate and live in frustration?

Why do we feel motivated – or beyond motivated,

even invincible, like we're sure things and realizing our hopes and dreams is a foregone conclusion – when we're at conferences, seminars, or other meetings of like-minded individuals, yet when we get home, those feelings of motivation, confidence, and self-assuredness evaporate within days?

Knowing what we know now, one could argue that it has to do with vagal tone and increased feelings of safety in those environments.

As I mentioned previously, one of my favorite books is *The Rise of Superman* by Steven Kotler and Jamie Wheal. In that book, the authors explore the role of flow states and how they have led to exponential increases in human performance, especially in extreme sports (surfing fifty-foot waves or climbing sheer rock faces without harnesses), where the athletes are literally forced to flow or die.

Aside from natural wonders and extreme risk, one of the most powerful flow triggers examined in the book is the phenomenon of "group flow." Kotler and Wheal point out that small communities of rock climbers, bands of outcast surfers, or skate park friends often push each other to successes beyond what they could have achieved alone.

Together they achieve group flow within sessions, but they also inspire each other and literally break barriers together. It's like the old McDonald's commercials with Larry Byrd and Michael Jordan playing horse, constantly one-upping each other. Or for a more realistic example, the way YouTube has allowed extreme sports athletes to document and share their groundbreaking feats, inspiring and fueling other athletes to push for longer jumps, bigger waves, or more rotations.

It's a real-life version of that Jordan-Byrd commercial, "Anything you can do, I can do better."

Consider any potted plant. Like humans, a plant's potential is restricted by the pot in which it grows. Once a plant reaches the maximum size that can be supported by the vessel in which it is potted, its growth is halted. Move the plant to a larger vessel, and its growth will resume – but only until it reaches the capacity of this new pot.

A plant can only grow as big as the pot that it is planted in. And like plants, our growth potential as humans is restricted by the box we keep ourselves in.

We, as people, can only grow to the constraints of the belief systems, societal norms, and the community in which we keep ourselves.

Expand your comfort zone, be mindful of your community, and be mindful of the (potentially limiting) belief systems that frame your subconscious and conscious thoughts, as they will either encourage or inhibit your growth as a human.

Look no further than the four-minute mile. For years, running a sub-four-minute mile seemed inhuman and unattainable. But once Roger Banister cracked that threshold in 1954, a total of sixteen runners accomplished the sub-four-minute mile in the following three years.

They just needed to know it was physically possible. Once that self-constructed "barrier" of human potential was removed, accomplishing the feat was not a matter of *if*, but *when*.

This shows the power of the mind and what we perceive as limitations on human potential. As soon as we know something is possible, we greatly increase our odds of achieving that feat.

To truly begin to live without limits, look for that four-minute mile example, that "Roger Bannister moment" in your life.

What perceived limitations have been holding you back? Whatever you can dream up, whatever you could possibly ever attempt, if you can imagine it, it's probably possible.

You simply need to step outside of that confining little pot you're planted in. We need to train our conscious minds to expand their framework and to get outside of our growth-limiting communities and beliefs.

Go. Do. Kick ass.

As *The Rise of Superman* illustrates, we see an inordinate amount of success coming from these small, tight-knit communities.

I postulate an alternative reason for the success and meteoric rise of the individuals within these communities and the industries that unite them, and that is the overlooked element of emotional safety that is supporting or even enhancing the creativity, drive, and confidence of the members of these communities.

Think about the things that you most want to go after – the thing that you really want to pursue, the place in your life where your feelings f*ck you up the most. Meditate on the places they hold you back, whether it's starting on a diet, learning how to cook for yourself, starting a new workout regimen, writing a book, or starting a business.

Whatever that thing is – the one that has been in the back of your mind as you've been reading this book, the pursuit in your life to which you have already begun thinking about how you can apply this book – I want you to think about that vision and answer these questions:

Do you feel safe pursuing that with everything that you have?

Do you feel like you will be negatively judged by your community, by your social circle, by the people in your life like your friends, your family, your co-workers, your boss, your peers?

Or do you feel like you have their full support? How comfortable are you flying your freak flag around them and openly talking about your passions, hopes, and dreams? Do you feel like they'll say, "Let's do this! How can I help?"

Chances are your world is the former rather than the latter. And if you're living in the former, your defense mechanisms will always be firing because your Vagus nerve knows it faces ostracization from your tribe if you pursue that thing, and that's why you're having the negative, limiting feelings that you're having.

Conversely, if you're in a society that supports such ventures, then your Vagus nerve will feel safe. It will sense that you can pursue these things, that you can venture out and do these things without having to fear judgment or being ostracized for being different, and that's why community is so important. That's why your tribe is so important when it comes to feeling safe, increasing vagal tone, and promoting emotional resiliency.

We know that safety with a tribe of close peers brings increases vagal tone, and with the corresponding inhibition of defense responses, we're free to operate at our highest levels – to create, to dream, to try new things – and this is ultimately why innovation is so commonly driven from small groups of "outliers."

I'm sure many such communities (business incubators, co-working spaces, etc.) exist throughout the world, but I'll focus on two examples here: Silicon Valley (because most people are familiar with it) and Venice, California (specifically, the community at Deuce Gym that, per capita, may have the highest population of successful people of any public access community I've personal encountered).

What makes these communities hotbeds for success? Is it the networking and insider connections? They surely don't hurt, but they also don't create success. They simply enable and facilitate the movement of something that already exists.

You can have all the connections in the world and get meetings with industry movers and shakers, but they mean nothing if you don't create the product, make the climb, or otherwise do the work that will impact others.

Is it the motivation that comes from being surrounded by other high-achieving individuals? Maybe. But maybe not – for every person motivated by such an environment, there is one frustrated and shamed by feelings of inadequacy. And even those who are motivated still must go home and actually *do* something of value with that motivation.

I asked leading high-pressure performance coach Andy Murphy about this very issue. Here's how he explained it: "If we have a low perceived self-worth, if we come from a fear response —that fight or flight default — even when we hang around high performers, all it's going to do is create self-doubt and negativity, and we're going to feel bad about ourselves. The default underlying emotion is the thing that affects our filter, our neurons, and our beliefs."

So then it must be the inspiration and ideation that comes from being immersed in such a community, right? Again, no. Everyone has ideas and inspiration, including your mother, the stock-boy at the local big box store, and you. It's the acting on those ideas that changes the world (and your life).

I'm arguing that while certainly beneficial, none of those things are the single biggest factor contributing to the disproportionate amount of success we see from such communities.

The reason for the success of the individuals involved them *feeling safe*.

Not what you were expecting, was it? Me neither.

Truth is, I've been to Venice and Deuce Gym for two years and never listed safety among my reasons for going there.

But that changed when I started researching the polyvagal theory and learned about the neurovisceral integration of feelings, emotions, and decisions.

As we've learned, the safer we feel, the greater our vagal tone.

The higher our vagal tone, measured by HRV, the greater our capacity to think, move, create, or otherwise express that which ignites the fires of our souls

The neuroscience research agrees: "Higher resting HRV is associated with more adaptive and functional top-down and bottom-up cognitive modulation of emotional stimuli, which may facilitate effective emotion regulation. Conversely, lower resting HRV is associated with hyper-vigilant and maladaptive cognitive responses to emotional stimuli, which may impede emotion regulation." [95]

This is why outcast bands of misfits throughout history – be they surfers, climbers, mystics, IT gurus in Silicon Valley, or Arnold-era bodybuilders in muscle beach – experience seemingly otherworldly growth and success.

To quote my friend Todd White, founder of Dry Farm Wines, "Darkness cannot exist in the presence of light. Self-doubt, fear, and frustration cannot exist in the presence of love, gratitude and abundance."

The people in these communities simply experience less doubt, fear, or otherwise limiting feelings and thoughts.

It's important to note that those potential downsides do not exist any less for them than those of us outside the pales of such communities, but because they feel safe in their tribes, their vagal tone is higher, their defense responses are turned off, so they *experience* less of the negative emotions, feelings, and thoughts that get in our way.

Understanding this is massively powerful for us. We can learn to recreate (or create) such defense response-inhibiting environments in our own lives, so that we can enjoy the same unbounded realities as if we were living fully immersed in such utopian tribes of like-minded individuals.

Obviously, we should seek to create these social circles whenever possible, but that may not be realistic for everyone or all the time. The rest of this book will be dedicated to showing you exactly how you can create this increased vagal tone to literally disable the governor that has been holding back your mind.

Incidentally, the conditions cultivated by these outlier communities are the same conditions created at Tony Robbins seminars and other "self-help" conferences around the world in order to foster those aforementioned feelings of invincibility.

They all recognize that feeling safe is a biological pre-requisite for experiencing our highest expressions of self. Understanding this and being able to favorably "hack" vagal tone can help us tap into these states of enhanced creativity, productivity, and performance.

Activities

Play

As Dr. Porges explained when I interviewed him, few things turn off our defenses like play and gratitude. Let's start with play. As I experienced when dunking on that eight-foot basketball court, play can be transformative for adults. It's nearly impossible to feel threatened and simultaneously be fully engaged in play.

Charlie Hoehn, author of *Play It Away: The Workaholics Cure for Anxiety* reminds us that "the opposite of play is not work. The opposite of play is depression." [96]

Throughout *F*ck Your Feelings*, we're talking about elevating our physiological states in an effort to positively impact our emotions, feelings, thoughts, and decisions. This is precisely what play does – both in the immediate present and in the long term.

As Hoehn points out, the opposite of depression (and anxiety) is play. Not coincidentally, most play involves movement, another powerful tool to alter our physiology for the better.

According to Stephen Porges, play is the epitome of safety. We would never truly play in the presence of a dangerous threat, so the act of playing is a signal to our sympathetic and limbic defenses to let down their guard, thereby increasing vagal tone. [82]

Psychology legend Abraham Maslow and the original guru of flow Mihaly Csikszentmihalyi have both extolled the virtues of play, specifically that they require full presence for genuine participation and "optimal experiences." [97, 98]

With this increased presence, our mind shifts to the now, shutting down the defense response of the limbic system, increasing vagal activation, and releasing a cascade of "feel good" neurotransmitters and hormones like oxytocin, endorphins, dopamine, and serotonin. [99, 100]

Play doesn't just make us happier and more resilient. As recent studies have demonstrated, a lack of play decreases our happiness. Children who play the least have higher levels of depression. In adults, a lack of recreational activity has been linked to criminal behavior (I'd call that poor decision-making), obesity, and reduced creativity. [101, 102]

A 2005 review of studies that investigated happiness and decision-making concluded that increased happiness leads to better decisions across multiple life disciplines, including health, business, income, performance, and relationships. [103]

As you can see, play is crucial. It positively shifts our physiology and mental states in the "now" and contributes to beneficial traits for long-term health and well-being.

So how can you get more play in your life?

From personal experience, I have found that solo play keeps me in my head and increases my likelihood of feeling guilty or continuing to think about daily life things. Adding a friend to the mix changes this for me. I'm forced out of my head and into the moment. My advice is to do your best to include a friend or friends – the oxytocin and socialization stacked on top of play will pay dividends.

Of course, I work alone, so I may crave interaction more than others. Maybe you work with a large group of people and would rather play without anyone else around. That's great too. Find what works for you.

<u>For altered states</u>

Drop everything and go play! It's crazy that as adults we have to remember to go play and even *how* to do it. (But I'm right there with you).

The key is to keep it unstructured and freely chosen. Struggling to come up with an activity?

Charlie Hoehn suggests asking yourself this question: "What would the ten-year-old version of you do right now?" Think back to the activities you enjoyed as a child; those are who you really are. Now, go do them.

Better yet, try to maintain a ten-year-old's enthusiasm and perspective in every minute of every day. Get down on the floor and roll around with a niece, nephew, grandchild, or pet. Joke, smile, laugh more, and ask questions. Learn, dream, play, and enjoy your time here. Stop being such a rigid adult.

Laughter is too often overlooked. Go see a stand-up comedian, watch or listen to comedy, read some jokes. Smiling changes our physiology, and laughter does to an even greater extent.

The saying goes, "Laughter is the best medicine,"

and as research shows, "outward expressions can amplify our internal feelings." So go ahead, laugh it up and increase your happiness. [104]

<u>For positively altered traits</u>

Ditch the digital fantasy football and join an actual flag football league. It counts quadruple because it's a) play b) outside in nature c) human socialization/ interaction and d) movement.

Not into football? No big deal. Swap out flag football for hiking clubs, basketball, surfing, improv class, bingo (no movement), or take Charlie's advice: think of the activities you enjoyed as a kid. Those are the things that will bring you the most joy as an adult.

Block off thirty to ninety minutes a week to play. This should be separate from your structured exercise regimen. If you're counting reps or logging times, you're missing the point of play. Relax, get lost, and have fun.

Turn Off Your Defense Mechanisms

Other than play, Porges offers the following advice to increase vagal tone.

First and foremost, it's all about turning off the defense mechanisms. We must feel safe. Rather, we must convince our vagus nerve that it is safe. Porges is careful to point out that the removal of threat is not the same as safety.

In today's non-stop society, it's important that we realize this. In most instances, we will not be able to isolate ourselves from things that can trigger our defense responses. So it should be our focus then to train our minds, control our responses, and actively seek activities that move us from the sympathetic state to the parasympathetic state.

Second, Porges recommends playing wind instruments. These instruments require breathe control, and this controlled breathing helps improve vagal tone. We'll talk more very soon about how breathe control can help us control and train the vagus nerve, prefrontal cortex, and limbic system.

Third, Porges encourages all humans to engage in *real* social interaction. Whether we consider ourselves introverts or extroverts, we're all mammals, and as mammals, we actually require the presence of others.

From the safety in numbers, the oxytocin releases and the accountability to our tribal peers, we have much to gain from person to person interaction. To that end, most high performers or successful people often credit their success to building support networks or systems. We did, after all, evolve in tribes, helping each other meet our needs.

The old saying, "It takes a village," may be applicable in this instance. Porges says this is crucial for improving vagal tone, stressing that we must "invite it and make it multi pathway. It must be give and take. The reciprocal loop makes it work."

Porges has actually given talks (you can find the videos on YouTube) detailing how ancient practices of group chanting, prayer, song, or even yoga/breathing exercises increase vagal tone.

All of these practices can work when performed alone, but the effects are compounded when performed with others.

Finally, Porges stresses the importance of feeling safe in the arms of another human. Porges is big on safety – and for good reason; making the vagus nerve feel safe prevents it from activating our defense responses.

Practicing gratitude can improve vagal tone as well, since gratitude in itself is a state of safety.

Gratitude

There are two things we always have control of: our attitude and our effort. Regardless of our chronological age, our brains are 200-million-year-old brains. Our bodies are carrying around the same tissue that has kept our species alive and allowed us to flourish throughout our time on this planet.

We haven't always been on top of the food chain. Part of our ability to survive, evolve and become the apex species on this planet required being on constant lookout for threats. Our 200-million-year-old brain is wired to look for danger and trouble, or in other words, we're wired to look for what's wrong. And we're really good at it. It's part of our superior pattern-recognition abilities discussed earlier.

But this is a process we no longer need as much as we did when our species was moving from cave to cave, eluding predatory saber-toothed tigers and group-hunting wooly mammoths for food, clothing, and other necessities.

Since this ability is wired into our brains, it's tough to turn it off. It's easy for us to see what's wrong in our lives and to complain. The problem is, as Tony Robbins reminds us, "There is always something wrong. If we stay in the conditioned mindset to look for what's wrong, we can always find something."

Ergo, if we allow ourselves to stay in that default pattern, enabling our limbic system to run the show, making us operate from a default of fear and survival, we are severely limiting our ability to expand as humans.

Some trepidation and risk management is good, like when buying a house or selecting a spouse. What we want to avoid is letting that pathway become our subconscious, default setting that runs the show without us even being aware of it.

We can follow the advice of Tony Robbins, Taoists and Buddhists, and train our minds to look for what is right. There is also *always* something good, like the fact that you woke up today, that you have eyesight to read this (or hearing to listen to this), that you have food to eat, a bed to sleep in, and a roof over your head.

As Stephen Porges taught us, gratitude is a safety practice that instantly increases vagal tone. Tony Robbins likes to call it "reframing," a practice where we shift our perspective on a certain experience or scenario.

Things do not have inherent meaning. Things, be they material things or worldly encounters and experiences, only have the meanings which we assign to them. Being aware of this can help us frame (or reframe) things to facilitate more desirable moods, feelings, and decisions. It's life-altering to be able to identify these feelings in real time and control them so that we can make rational, logical decisions that are in line with our life's true purpose – our soul, spirit, or self, instead of our ego or emotional feelings.

When you're in a beautiful state, you do the right thing for yourself, your family, and everyone else. — Tony Robbins

Reframing is one tool that we can use to get to that beautiful state so we can make the right decisions, audit these feelings, and guide our decision-making process in real time. [105]

Resist temptation to act quickly based on emotion. Realize it is an emotional thought, accept that, let it pass, and use your prefrontal cortex to make decisions that are in line with your long-term goals, not your emotional state.

One of the easiest, most profound, and most research-backed ways to shift ourselves into that beautiful state, as Robbins calls it, is to practice gratitude.

My hypothesis is that gratitude, as Porges says, increases HRV by helping us turn off our defenses and feel safe, which results in higher levels of energy and greater physical expressions of "health." [106, 107]

Gratitude researcher Robert Emmons has published a handful of papers showing the positive impact of gratitude on mental health. Most notably, frequent gratitude practices reduce instances of envy and depression and increase happiness. [108]

Researchers at the University of Kentucky have found that gratitude increases empathy and reduces aggression. [5,6] Gratitude practices have also been shown to improve sleep, which is one of the single biggest factors in increasing HRV. [109, 110, 111]

How can we build this into our daily routine

then? The two most obvious ways are to incorporate gratitude into our morning and evening routines. You could also do a gratitude reset as part of your pre-meal blessing or prayer.

Incidentally, love and gratitude are the two words that had the most profound impact on crystal formation of water, making them good things to say before eating and drinking anything, perhaps increasing the nourishment they deliver to us. [112]

Starting your day by taking two minutes to consciously thank someone or something for another day alive is a powerful way to start your day.

Journaling at night is also a good option for gratitude practices, as the study mentioned above found that gratitude practices before bed improved subjects' quality of sleep, making it a powerful synergistic combination to increase HRV. [113]

There is no wrong time to be grateful. There is, however, a wrong way. According to research presented in Adam Grant's *Originals*, people asked to list three things they loved about their life found this task easy and noticed an improvement in mood. However, people asked to list twelve things found themselves grasping for items to complete their list, making them feel as if they did not have many things for which to be grateful. This left them filled with unintended negative emotions that were opposite of the intended effects. [114]

Keep your gratitude practice simple. Start small and list one thing, like being alive, or having your eyesight, or having food security. I suggest listing no more than three items per day and trying not to repeat an item every day. (Listing the same three items every day will make the practice lose its intended effect).

Also, think about who you can help. People who express the most gratitude also tend to be the most giving, [115] so while you're counting your own blessings, think about who else you can help.

When we stop focusing on the problem, distract our minds, and think about abundance, hope, gratefulness, and charity, we change our physiology for the better. Our mood is different, and we're in a better place from which to make decisions that will positively shape our lives.

Porges likes gratitude for its ability to shut down our defense response and increase vagal tone, which increases emotional resiliency both long term and in the moment.

Tony Robbins likes gratitude for its ability to move us into that "beautiful state" where we always make the right decision for ourselves and our families.

Science likes gratitude because it demonstrates effective, measurable results.

If you're not already practicing it in some form or fashion, you'll love gratitude because it is a simple, easy to implement, incredibly powerful habit you can add to your daily routine in less than five minutes a day for immense benefits.

If we look closely, each of Porges's recommendations for increasing vagal tone resembles every stress reduction protocol we've ever heard: play more, practice gratitude, increase real social interaction, help and give to others, meditate, and learn to breathe properly.

When in doubt, spend more time in parasympathetic states.

Measuring HRV

Measuring HRV requires both a hardware device and a software component, usually an app on your phone. There are a number of apps and measuring devices, and while I've tried many of each, none are as friction-free as I'd like them to be.

In fact, they're such a hassle that I don't measure my HRV daily.

With traditional heart rate straps, you need a Bluetooth connection to an app on your phone. When you wake up in the morning, you have to get the heart rate strap on, and you have to link that to an app on your phone that can tell you what your HRV is.

It is kind of a pain in the butt, because you have to get up, get the electrodes wet, stick them on you somehow, get the Bluetooth to sync, and then get a measurement.

To me, that just defeats the purpose. If I am up, then I am up. I am not getting back into bed, I am not going to lay down to take that reading, so how do you figure out a way to get that thing on you without getting up and out of bed?

I was really excited when this new company called ithlete came out with a finger sensor that measures your pulse on your finger and then plugs into the audio jack of your phone, but unfortunately, that one wasn't any better. Some days it would give a reading the right way, other days I would sit there for twenty minutes and never get a reading, so that proved to be equally frustrating, despite my excitement.

Because I have not found a way to measure it without frustration, I have stopped measuring it on a daily basis.

One of the fastest-growing fitness trackers is the Oura ring, and for a while, they have been recording what they call a "readiness score." I recently spoke with a high-level executive at Oura who told me that they now have the capability to report true HRV, and that their readiness score is actually a composite of the HRV and some other metrics. Another benefit of the ring versus a hardware device that you apply in the morning is that the ring gives you 24/7 HRV data, so you can look at it straight when you wake up in the morning, but you can also check it at midday and in the evening – it is all right there on the app.

Not only is it convenient, but you are also getting it all day, every day, and not just that instantaneous moment where you strap something on and you get a reading.

If I have to give a recommendation, then at the time of this writing, the Oura ring is the best tool for monitoring HRV that I am aware of, especially since it reports 24-hour data as opposed to a single snapshot in time like a chest strap or finger sensor.

As HRV technology improves, I predict it will become an increasingly more common modality to measure overall health and emotional health, and continue to help us balance the stressors from our training loads.

CHAPTER 5:
The Gut – Our Second (or Maybe First) Brain

The Gut: Our Second (or Maybe First) Brain

With all this talk of the brain impacting the gut and vice versa, it's time we explore the neurovisceral integration from another angle – the gut as our second (or maybe our first) brain.

Paleoanthropologists and evolutionary researchers believe our ancestors, like most other species, had to rely on their digestive systems for information, direction, and guidance. Modern neuroscientists and gastroenterologists are drawing similar conclusions from their research.

To understand how this is possible, we must look at how the human nervous system develops. During our nine-month gestation period in our mothers' wombs, our nervous system is among the first things to develop.

This development begins with something called the enteric nervous system. Formerly classified as part of the autonomic nervous system, the enteric nervous system consists of millions of neurons that govern the function of the gastrointestinal system and has recently been classified as its own nervous system since it contains an independent neural activity. [116]

During gestation, our enteric nervous system gives rise to the central nervous system, our brain and our spinal cord, making the gut-brain connection impossible to ignore and ridiculous to downplay, as the gut literally gives birth to the brain. [117]

The brain and visceral organs, as we have previously discussed, are connected via the vagus nerve, and this link is extremely primitive and rooted in survival.

If we go back to the domestication of fire and the beginnings of homo sapiens, we see that fire enabled cooking. As paleoanthropologist Pascal Picq explains, "cooking is a form of pre-digestion, making digestion and assimilation of nutrients easier." According to Picq, "cooking foods provides 16X more energy." [118]

What does this have to do with feelings, choices, and decision-making? A lot, actually.

Cooking as a form of pre-digestion, and the increased assimilation of nutrients, was a huge evolutionary advantage for our species. This decreased our required investment in resources for digestion and allowed our upper brains (the central nervous system) to develop, as we did not have to devote the entirety of our resources to digestion and the subsequent acquisition of more food.

Let's take a brief pause from our own brains and feelings to think about Maslow's hierarchy of basic psychological needs: food, water, rest, shelter, and security. [119]

Most every animal on the planet spends the majority of its existence in pursuit of food, shelter, or reproduction. If you're thinking of a counter argument (like I did) that involves any of the societal constructs in the animal kingdom we see on the Nature or National Geographic channels, remember that those constructs are put in place to decide pecking orders for Maslow's basic needs – distribution of food, procreation, shelter, physical health, and continuation of the species.

Since we humans are the only organisms on the planet that cook their food, it's hard to argue with Picq's line of thinking that this dietary change played a massive role in our mental evolution. The highest expression of human existence is to have these needs met, so we can turn our focus to the education, growth, contribution, creation, innovation, and love that bring fulfillment and satisfaction to our lives.

From an evolutionary standpoint, this adaptation decentralized command and allowed our upper brain to evolve to eventually perform the higher-level tasks that we, as humans, can perform that other mammals cannot – things like inventing cars, planes, the Internet, and smartphones and sending satellites into space.

This was part of the evolution of our brain (and our species), helping us jump physiological hurdles and laying the foundations that, over the following hundred thousands of years, led us to the species we are today.

From this perspective, our gut is not a second brain, but actually our first brain – albeit a primitive one.

Gut Neurons and Bacteria

We mentioned that the enteric nervous system is comprised of millions of neurons (somewhere between 100 and 500 million) and contains enormous amounts of bacteria.

To help contextualize the scope of the gut's sentience, those 100-500 millions of neurons are about 5X as many neurons as we have in our spinal cord and 2/3 the number of neurons in a cat's entire nervous system. [120, 121]

Which begs the question: if cats and dogs are sentient, intelligent organisms, then so is our gut... right?

All neurons, be they in our brains, spinal cords, or guts, communicate via chemical messengers called neurotransmitters, of which scientists have discovered more than 100 types. [122] We'll focus on a few major neurotransmitters in just a minute – serotonin, GABA, dopamine, acetylcholine, and anandamide.

We also know that we have an equal amount of bacteria inside us as there are human cells. (This number is often misquoted as 100-1000X more bacteria cells, but researchers debunked and clarified that myth in 2016). Still, this makes our microbiome the most concentrated ecosystem on the planet. This microbiome is also called the gut biome, or gut microbiota, and I'll be using those terms interchangeably. [123]

In short, it's difficult to overstate just how significantly our microbiome influences the communication between our gut and brain.

These bacteria greatly impact our digestion and absorption of nutrients from the foods we eat, which in turns impacts energy metabolism and the subsequent resources available for immune function, cardiorespiratory function, locomotion, and cognition.

Speaking of cognition and mental performance, our microbiome also influences (and in some instances creates) the chemical messengers, aka neurotransmitters, that carry our nervous system communications across neural synapses.

These neurotransmitters regulate appetite,

cravings, mood, feelings, fears, stress response and balance, memory, and neural plasticity. If we zoom back out to our 50,000-foot view, this impacts decisions and can be even be measured with vagal tone. [124, 125, 126, 127]

Gut Function/Dysfunction and HRV

All disease begins in the gut. – Hippocrates

Hippocrates, the father of modern medicine, was right. Our microbiome and neurotransmitters are part of an intricate dance between gut and brain and are inextricably linked to overall health, vagal tone, and cognitive function – especially as it relates to emotions, feelings and mood.

We can measure the health of this relationship with HRV, which also gives us insight into the health of our emotional resiliency/bandwidth/ability to stiff-arm those negative emotions, feelings, and thoughts that can derail us.

As we look at this research, keep in mind that our gut is connected to the brain via the vagus nerve, and vagal tone is the measure of overall well-being, emotional bandwidth, and sympathetic/parasympathetic balance.

Since this neurovisceral system is connected by the vagus nerve, the polyvagal theory governs the hierarchy of its actions and is its default operating system. Therefore, all of the training we can do to increase prefrontal cortex control over the limbic system applies to increasing emotional well-being, as well as gut health and anything else linked to HRV.

The link between gut dysbiosis as well as gut

inflammation and our emotional resiliency – as measured by HRV – has been verified by multiple studies and is currently being investigated with great excitement in the scientific community.

Let's start with one eye-opening study from 2016 that realizes the "changes in sympathetic/parasympathetic balance regulated by the brain precede changes in the inflammatory cascade" which leads to "the pathogenesis of obesity and diabetes mellitus and its complications, as dysbiosis is thought to play a pivotal role in diabetic-associated disorders." [128]

These researchers were investigating the link between gut inflammation/dysfunction and how it may precipitate obesity and diabetes, when they noticed that negative changes in vagal tone precede the inflammatory cascade. Interestingly, this inflammatory cascade involves the same anticholinergic pathway we'll discuss in the upcoming acetylcholine section.

A report published by *Harvard Health* demonstrated that decreased HRV, or elevated stress (unbalanced sympathetic state activation), is linked to gut dysfunction, stating that "stomach or intestinal distress can be the cause or the product of anxiety, stress, or depression. That's because the brain and the gastrointestinal (GI) system are intimately connected." The review of thirteen studies concluded that "patients who tried psychologically based approaches had greater improvement in their digestive symptoms compared with patients who received only conventional medical treatment." [129]

The link is undeniable, although not yet fully understood. However, we can apply this knowledge and say that all of the training we can do to increase prefrontal cortex control over the limbic system applies to increasing emotional well-being as well as gut health and anything else linked to HRV.

Research continues to further investigate these links, and I suspect that vagal tone will become an increasingly unique and important biomarker to both track and train via HRV and parasympathetic activities respectively.

How do you do this? We can start by increasing awareness, elevating consciousness, and helping our primitive bodies move away from defense responses and spend more time in the restorative parasympathetic state.

No discussion of the gut-brain connection is complete without an examination of the language they use to communicate: neurotransmission.

The Chemicals in Our Brain (and Gut)

The mind is its own place, and in itself can make a heaven of hell, a hell of heaven. – John Milton, Paradise Lost (1667)

When Milton penned these words in his iconic 1667 poem *Paradise Lost*, he brilliantly captured the duality of our minds.

Get it right, and we're unstoppable forces of nature, capable of anything we can imagine. Get it wrong, and we're self-sabotaging piles of frustration and stagnation.

Whether Milton knew the actual mechanisms behind our thoughts and feelings is irrelevant. And whether we like it or not, our experience of the world is inextricably linked to the ecosystem in our gut and the subsequent physiological environment inside our heads.

That acute environment in our brains on any given day is greatly influenced by the chemicals in our brains. I'm not talking about the chemicals with the Green Mr. Yuck sticker on the bottle.

I'm talking about neurotransmitters, the chemical messengers that transport communications across the synaptic clefts (junctions) between nerve cells, aka neurons, hence neuro-transmission. Our nervous system is the communication highway in our bodies and without these messengers, our bodies would not be able to communicate.

Enter Michael Gershon, referred to as the "Leonardo of the Bowels," an expert in the field of neurogastroenterology and author of the 1998 book *The Second Brain*.

He also serves as the chairman of the Department of Anatomy and Cell Biology at Columbia University Medical Center. According to Gershon, "Neurotransmission is the language that nerve cells use to talk to one another. Neurotransmitters are the words."

Our ability to resist depression, feel happy, and think clearly is severely influenced by messages sent to the brain by the gut through these neurotransmitters.

There are more than thirty recognized neurotransmitters, but we're going to focus on four for now: dopamine, serotonin, acetylcholine, and GABA.

As we discuss the neurotransmitters, it's important to keep in mind acute and chronic settings. Just as weather is now and climate is long-term, we need to differentiate between acute, short-term, transient states and their climate-like partners: traits.

Before we dive deep into each neurotransmitter and how it impacts our feelings and decisions, here is a quick primer on their respective roles:

- Dopamine is known for motivation, drive, vim, vigor, confidence, and the desire to get things done.
- Serotonin is the feel-good neurotransmitter

responsible for positive mood, and it keeps fear, anxiety, and cravings in check.
- Acetylcholine is linked to mental-processing speed, lateral thinking, and creativity.
- GABA is the inhibitory neurotransmitter. Unlike the other neurotransmitters, its role is neural inhibition – it is the brakes for our neural network and is associated with feelings of calm and relaxation.

Dopamine

Fewer than 1% of the 100-billion neurons in our brain can generate dopamine, the neurotransmitter linked to motivation, drive, pleasure and reward, balance, and learning. [130, 131]

For years, scientists have theorized that dopamine's main role is in pleasure and reward, but our understanding of dopamine has been (and still is) undergoing a massive shift in recent years.

New research on this neurotransmitter is challenging the long-held belief that action creates reward and thus triggers the release of dopamine, perpetuating the cycle.

In fact, the shift in the scientific community is not complete, as many still talk about dopamine as the "do that again" signal at the end of the pleasure-reward circuit. Dr. Michael Merzenich, the father of neuroplasticity and founder of BrainHQ, called dopamine "the replay button" when he was on the Optimal Performance Podcast in 2016. [132]

He continued: "Dopamine is released to tell us 'that was good, do it again'" – hence the "replay button" moniker.

But recent research is challenging this long-held belief with new findings that suggest dopamine is involved on the front of this motivation cycle rather than as the reward on the back end. The research shows dopamine may be more involved with the motivation and initiative to begin and the drive, or perseverance, to continue than with the pleasure/reward for having done.

One of the earliest stirrings of this shift in understanding of dopamine came when a 2005 review of dopamine's role in pleasure and reward concluded that "the role of dopamine in the reward process was classically associated with the ability to experience pleasure; recent data suggest a more motivational role." [133]

My first exposure to this new paradigm shift came from a Society for Neuroscience article where they cited several mouse studies correlating a lack of dopamine to the reduced desire of mice to move or even eat. In some of these studies, mice with an inability to produce dopamine wouldn't even drag themselves across a cage to eat [134, 135].

This is powerful information for us to keep in mind as we seek to understand how our feelings drive our decisions. A dopamine imbalance – or in this case, extreme deficiency – caused mice to become so labile that they would not move across their cage for food.

Clearly, dopamine plays a role in decision-making and is something we should be aware of in our own minds.

There is more; a 2013 study published in the

neuroscience journal *Neuron* concluded that "the widespread belief that dopamine regulates pleasure could go down in history with the latest research results on the role of this neurotransmitter. Researchers have shown that regulates motivation, causing individuals to initiate and persevere to obtain something." [136]

John Salamone is a dopamine researcher and Professor of Psychology at the University of Connecticut who supports this theory, saying: "Low levels of dopamine make people and other animals less likely to work for things, so it has more to do with motivation and cost/benefit analyses than pleasure itself." [137]

The research is ongoing, but it's becoming pretty clear that dopamine is less of a "replay button" and more of spark to spur us forward. Dopamine powers the brain and drives our initiative to make things happen.

Is it the neurochemical that turns motivation into actions? And, if dopamine is such a force in regulating motivation, our initiative and drive, then what regulates dopamine levels and how do we keep them optimized?

Dopamine is, and must be, tightly regulated by our body, which is why only 1% of our 100-billion neural cells can generate it.

Here are two easy ways to keep dopamine levels from dipping too low: gratitude and morning sunlight. Gratitude stimulates the region of the brain responsible for dopamine, giving us a boost, but also explaining (in part) why happy, grateful people live longer than angry, bitter people. [138]

Daily exposure to sunlight, especially morning

sun, can support optimal levels of dopamine as well. It's crucial that this is retinal sun exposure for dopamine production, which means no sunglasses, as they block the sun's rays from reaching the retina. [139, 140, 141]

In addition to playing a major role in motivation and drive, dopamine has been closely linked to muscular coordination, movement, balance, and neuromuscular disorders, including multiple sclerosis and Parkinson's disease. Those are fascinating studies with meaningful applications in health and medicine, but they're outside the scope of this book.

Finally, dopamine plays a role in something called "pattern recognition" – a human ability that cognitive neuroscience describes as matching information from external stimuli with information retrieved from previous experiences or memories. It is our ability to connect the dots or link new ideas together. [142]

Dopamine, Time and Pattern Recognition

In 2014, Mark Mattson of the Department of Neuroscience at John Hopkins University published a paper in *Frontiers in Neuroscience* titled "Superior Pattern Processing is the Essence of the Evolved Human Brain," where he explains human pattern recognition as a function of our nervous system that has been cultivated and conserved through evolution.

According to Mattson, "superior pattern processing is the fundamental basis of most, if not all, unique features of the human brain including intelligence, language, imagination, invention, and the belief in imaginary entities such as ghosts and gods." [143]

Here is the summary of his words in layman's terms.

We're better than lower animals at processing visual and auditory patterns due to our more evolved, more robust neural networks. This enables us to maintain awareness of our place in our environment and to remember the locations of survival needs, things that are both good (food/shelter/water) and bad (predators and cliffs).

For more complex patterning that separate human intellect from lower animals, like imagination, inventions, and the creation of language, our visual cortex, the previously discussed prefrontal cortex, and the parietal-temporal-occipital expanded and evolved, enabling superior pattern processing.

This superior pattern processing may be the basis of most (if not all) higher functions of the human brain, including consciousness, language, mental fabrications (our constructs – religion for example), and time travel.

This superior pattern recognition is precisely why artificial intelligence is being built around machine learning that seeks to mimic human pattern recognition and pair it with superior computational and processing abilities of today's computers.

In his *New York Times* bestselling book *How to Create a Mind: The Secret of Human Thought Revealed*, MIT-schooled futurist Ray Kurzweil states that this "digital cortex" modeled after our human neocortex will process information 1,000 to 1,000,000 times faster than our human brain. [144]

We can use our powers of superior pattern

recognition in many ways to optimize our output and experience of the world if we take an objective look at patterns in our own lives.

What events, actions and scenarios inspire us and propel us to move closer to our stated goals? Identify and do *more* of these things.

What events, people, routines, etc. have the opposite effect? What patterns do we notice around the emotions, feelings, and thoughts that drain us, kill our confidence, and add stress to our lives? Identify and do *less* of these things.

In *The Rise of Superman*, Jamie Wheal and Steven Kotler examine the neuroscience of flow states and how these optimal states of consciousness enable exponential advances in human performance, both physically and mentally.

In the book, they describe the neurochemical cocktail of flow in great detail, but as it relates to pattern recognition, we can glean this much from the current body of science:

"Specifically it appears that dopamine helps to increase focus and to reduce 'signal to noise' ratios, which in turn help us gather more information quickly. Norepinephrine [a metabolite of dopamine] also enhances focus while anandamide increases lateral thinking by increasing the base of reference so that you are more likely to identify unique connections that you might otherwise have missed."

It's a self-fulfilling cycle of sorts, and it hinges on dopamine function. Dopamine helps to increase focus, which facilitates information intake, which helps pattern recognition – which, like working memory, is regulated by dopamine.

Increasing Performance Through Enhanced Superior Pattern-Processing Abilities

Since we know how the system works, we can take steps to optimize it and use it to our advantage. We can train our pattern recognition pathways and enhance the brain traits involved to drive improved performance in many areas of life; from survival to thriving, pattern recognition is major factor in nature – hunting, surfing, traffic, stock markets, and business trends to name a few.

Here's how to train it. Increasing superior pattern processing is a two-step process, and step one is increasing information intake.

At its core, pattern recognition is information processing matched with memory. By this definition, our own experiences and personal library form the maximum capacity of data points to which we have access.

According to Kurzwiel's book, a master in any given field has approximately 100,000 chunks of knowledge related to that field. So get to collecting those chunks! The more you learn and the more you do, the more potential dots you have to connect. You don't have to be a speed-reader or super learner, although pattern recognition is certainly a major application of those skills.

We all know that person on the opposite end of this spectrum – the person with limited life experience, singular perspective, and therefore a stunted ability to understand and relate to new concepts. Don't be that person. If you're reading, you're likely a growth-minded person, so nothing to worry about there.

There are some research-backed ways to improve the speed and accuracy with which we receive information and develop our pattern-recognition abilities.

Playing games and sports has been shown to increase pattern recognition, as exposure to patterns and scenarios wires our brain to see and anticipate certain repetitive events. Sports are little more than patterns, from playing catch with your dad in the backyard, to the pitch sequence in Major League Baseball or the play calling of the NFL. [145]

As we learned above, increasing focus during anything we do helps us take in more information. This is simply the act of being present and not being distracted or trying to multitask. Recall a time when you were extremely distracted and forgot something important or missed some environmental information – like talking on the phone while driving, or trying to eat or adjust the AC/heat and nearly missing your exit, a stop sign or a red light.

Neurofeedback, mindfulness, yoga, and meditation are all practices that increase our presence and reduce distractibility. Interestingly, many of the flow triggers outlined by Kotler and Wheal in *The Rise of Superman* can aid in this pursuit, as "flow follows focus."

Said another way, focus is a prerequisite for flow. And focus requires us to tune out distractions, something we are more prone to do in the rich environments detailed in the flow triggers – things that help us drive our attention to the now.

In case you're wondering, a few of these triggers include: high consequences, rich environments, proper challenge-to-skills ratio, immediate feedback, and sense of control.

The second step in this process to train and develop our pattern-processing ability is to improve our ability to access to this ever-growing database through something called "working memory."

Guess which neurotransmitter plays a role in working memory? Dopamine.

Brown University neuroscience researcher Himanshu Sharma defines working memory this way: "Working memory is the ability to keep our thought on information acquired in past, in light of the present demands, in order to plan for our future action. This process is mediated in our prefrontal cortex and requires dopaminergic transmission." [146]

As I told you earlier, everything is everything. Our body is a finely tuned machine, and almost nothing works in isolation, making health and self-care supremely important as every cell, tissue, organ, and organ system affects many downstream process of human life.

Here are two tricks to improve your working memory.

1) Dual N-Back

Dual N-Back brain training is a popular method of increasing memory and one used by many memory experts, super-learners, neuroplasticity training websites, and biohackers.

Multiple meta-analysis papers agree that Dual

N-Back training benefits cognition and IQ with as little as twenty minutes a day. [147, 148, 149]

This game-like system improves fluid intelligence and overall IQ by increasing working memory, which is considered to be the "bottleneck" of information processing. The greater this capacity becomes, the more information we can process and the faster we can do it.

Think about the card game Memory, where you tried to match cards that were laid face-down by remembering which ones were placed where. Dual N-Back is similar to this; except, as you get better, the training increases the number of data points between the two events or images you have to match. That number of "spacer" data points is the "n" that you have to go "back" in your memory.

You can find Dual N-Back apps for your phone or access it on a computer by searching "dual n-back."

2) Visualize Like a Memory Grand Master

Grand Master of Memory Mattias Ribbing suggests using visual imagery to increase memory, since our visual memory is greater than the language portion of our memory. His memory technique "Ribbing's Bike" involves placing items from a list onto specific parts of an imaginary bicycle to help with recall. For example, if the list includes fish, celery, and eggs, you would envision the fish spinning in the spoke of the front wheel, the celery as the handle bars, and the eggs sitting in the basket on the handle bars.

When you go to recall the items on your list at the

market, it's easy to recreate that visual image. Now you pick up the fish, celery, and eggs at the market and get on with your day.

Research from the University of Iowa supports this, as their 2014 study found that visual memory is significantly stronger that auditory memory. Tactile, or touch, is even stronger than visual memory, which may explain why so many of us prefer to "learn by doing."

If you can only hear the information coming in, follow Ribbing's advice to create a strong visual image of the information and imagine being able touch and move your image in three dimensions. Ribbing advises that the more you use this technique, the faster and stronger you'll become.

Serotonin

Known as the feel-good neurotransmitter, serotonin regulates mood, fear, and anxiety, and is associated with feelings of well-being.

Despite its profound impact on our mood and feelings, only about 5-10% of this neurotransmitter is produced in the brain itself. 90-95% of the feel-good neurotransmitter serotonin is produced in the gut by bacteria. [150]

Serotonin's impact on our neurovisceral integration can be divided into digestive and cognitive roles.

As "Leonardo of the Bowels" Dr. Gershon notes, "In the upper brain, serotonin means well-being, while in the gut, it regulates immunity and gut transit time." [151]

The cognitive role of serotonin is the one we most

often think of when we mention this neurotransmitter: well-being, regulation of fear and anxiety, and positive moods. As you would expect, this gut-brain connection is quantifiable through vagal tone measurements.

A 2016 meta-analysis of HRV studies and its association with dysfunctional mood confirmed substantial reductions in HRV across psychiatric disorders, and these effects remained significant even in medication-free individuals. [152]

The digestive role of serotonin is, as Dr. Gershon states, in immune function, gut transit time, and of course, is produced there by the beneficial bacteria in our microbiome. Proper functioning of this intricate system therefore relies on optimal gut microbiome. Dysbiosis, or maladaptive, imbalanced gut microbiomes negatively impact serotonin and mood in two ways.

First, the dysbiosis is associated with inflammation. It can cause inflammation and it can be caused by inflammation in the gut.

Second, the imbalance associated with dysbiosis can mean suboptimal levels of the beneficial bacteria that produce serotonin in the enterochromaffin cells of the gut lining. And that means disrupted (and dysfunctional) production of serotonin itself, which has downstream impacts on serotonin cognitive role in regulation of anxiety, fear and mood, as previous noted in the studies linking decreased HRV to disordered mood regulation.

In 2016, researchers demonstrated the ability to "transfer the blues" or depression from one organism to another, when they took the gut bacteria from depressed humans and used it to colonize the digestive systems of rats. As you might guess, the rats became depressed.

According to the researchers, "Fecal microbiota transplantation from depressed patients to microbiota-depleted rats can induce behavioral and physiological features characteristic of depression in the recipient animals." [153]

Given that selective serotonin re-uptake inhibitors, a class of anti-depressant drugs used to prevent the clearance of serotonin from the bloodstream in an effort to positive impact mood, are one of the most widely prescribed drugs in America, it should be no surprise that serotonin imbalances are linked to depression and other mood disorders.

In a paper titled "America's State of Mind," MedCo announced that "the number of Americans on medications used to treat psychological and behavioral disorders has substantially increased since 2001; more than one in five adults was on at least one of these medications in 2010, up 22 percent from ten years earlier."

In addition to serotonin's link to disordered moods, there is also new research showing a link back to disordered digestive functioning, including irritable bowel syndrome and gut permeability issues. [154, 155]

Dysfunction of this system creates a negative cycle that perpetuates this further crumbling of this finely tuned system. As gut health researcher Ferit Rahvanci explains, "The number one cause of Leaky Gut, IBS, and other gut disorders is inflammation. And the leading cause of inflammation is food intolerances." [156]

Diving further into this rabbit hole would tangentially explore the dietary component of gut health, which would be a book of its own, so I am being succinct on purpose.

Inflammation also reduces neuroplasticity and negatively impacts memory. Here's further reason to avoid inflammatory foods: they make us dumber. An Italian study from 2008 confirmed that increased systemic inflammation leads to a reduction in neural plasticity, the brain's ability to form new neural networks. [157] In 2014, UC Irvine neuroscientists Jennifer Czerniawski and John Guzowski wrote in *The Journal of Neuroscience* that "brain inflammation negatively impacts memory." [158]

Forming new neural networks is exactly what we're talking about in this book. New levels of awareness, training the brain to operate at higher levels, creating new habits for ourselves – these are all possible because of our brain's neuroplasticity.

There is no doubt that chronic inflammation is a killer of energy, mental performance, emotional resiliency, decision-making, and long-term health.

Beyond our day-to-day performance, inflammation has been linked as the root cause of seven of the top ten causes of death. The diseases of inflammation include Alzheimer's, Parkinson's, cancer, depression, diabetes, and other metabolic disorders, to name a few. [159]

Recognize the importance of the foods you put into your body and understand how they impact gut health, nutrient absorption, energy levels, mood, neurotransmitter production, cognitive performance, and immune system health. Protect your gut health by nourishing it with prebiotics and probiotics daily and by avoiding causes of inflammation like chemicals, pesticides, high-sugar foods.

Our body is an intricately woven interlinked series of systems, and as we continue to see, everything is interrelated.

German Professor Dr. Michael Schemmann of the University of Cologne agrees, noting that we carry emotional issues in our gut, creating a state of hypervigilance. He reminds us that irritable bowel syndrome and other gut dysfunctions are mostly stress-related (measured by low HRV scores). His research shows that irritable bowel syndrome stems from dysfunction of the communication between mucosal surfaces of the intestinal wall and the nerves in our gut, implicating both the vagus nerve and serotonin. [160, 161, 162]

The faulty communication results in increased neural firing and leads to hyperactivity and chronic inflammation.

He posits (and I agree) that drug intervention is not needed. We simply need to break the cycle of introducing offending foods (I found mine using a food intolerance test called Pinnertest), work to heal and cultivate a favorable microbiome, calm our enteric nervous system, and support natural serotonin production through the following methods.

Sunshine and clouds have long been used as analogies for happiness and sadness respectively for good reason. New research has found that our bodies produce less serotonin on cloudy days. In fact, the levels of serotonin in the brain were directly related to the duration of bright sunlight. [163, 164] So those feelings of being down or less motivated on cloudy, rainy days are legitimate – there is a physiological difference in our bodies and brains.

I'm not mentioning this to give you an excuse. Rather, I'm mentioning it so we can be aware of it and avoid it. I want to be in control of my physiology – not the weather, not any external circumstance. We'll list and discuss a myriad of physiological state "hacks" like movement, music, gratitude, and more in this book that we can use to overcome rainy days or any other barriers to our mission. More research is being conducted, but the link is obvious. If you're feeling down, spend more time outside in the sunshine.

Another report shows that we produce less serotonin when we become more sedentary, which is why movement is a big part our program. [165]

As the Mayo Clinic wrote in a 2000 study titled "Serotonin as Mediator of the Gut-Brain Connection," the gut and brain are physically connected via the vagus nerve, whose health is measured by HRV, making serotonin a crucial mediator of sympathetic/parasympathetic balance, overall health, and emotional resiliency. [166]

Acetylcholine

A combination of acetic acid and choline, acetylcholine is the neurotransmitter responsible for arousal, attention, memory, motivation, and mental processing speed. [167]

It's helpful to think of this neurotransmitter as a

neuromodulator, or the throttle that controls the speed at which our brain functions in any given moment.

Personally, I experience exceptional lateral thinking and connecting of ideas when I supplement with choline or choline-boosting nootropics.

Acetylcholine is also found in our neuromuscular junctions (where nerves connect to muscles) and plays a role in motor unit recruitment and movement. This is why L-Alpha glycerylphosphorylcholine, or Alpha-GPC (a form of choline) is often used as pre-workout to facilitate increased motor unit recruitment, experienced as faster and more explosive muscle contractions. A 600mg dose ninety minutes before weight training has also been shown to increase serum levels of growth hormone post-workout. [168]

That's fun if you're into weight room results, but it's a little off topic. The acetylcholine and vagus nerve connection involves not muscular function, but immune function.

A study published in the Journal of Internal Medicine in 2015 titled "The Pulse of Inflammation: Heart Rate Variability, the Cholinergic Anti-Inflammatory Pathway, and Implications for Therapy", is the exact link we're looking for.

In that paper, the researchers summarized their findings as such:

"Numerous studies have investigated the relationship between depression, systemic cytokine production, and heart rate variability. Depression is associated with abnormalities in innate and adaptive immune function, including increased production of pro-inflammatory cytokines, decreased production of anti-inflammatory cytokines, and increased expression of surface markers associated with immune cell activation. It is plausible that over-expression of cytokines in the brain may influence depressive behavior, because cognitive impairment, behavioral dysfunction, and sickness syndrome effects are mediated by cytokines, including TNF. Current data are unable to determine whether the onset of a major depressive episode precedes the development of a dysfunctional immune response, or vice versa. Patients with major depressive disorder also exhibit decreased heart rate variability, and the severity of the impairment correlates with the clinical severity of depression". [169]

Here's what that means for us: cytokines are a broad category of cells that act as immunomodulating agents. Elevated levels of the specific inflammatory cytokines mentioned in this study, interleukins and tumor necrosis factor, are signs of increased inflammation in the body. [170]

As this study reiterates, inflammation accelerates (if not causes) many of the top causes of death in the United States today, including heart disease, stroke, cancer, diabetes, and sepsis.

Increased inflammation is linked to decreased HRV, which makes sense, as increased inflammation is a signal for the body to fight inflammation, triggering our sympathetic fight or flight response. This is another mechanism through which low-level chronic inflammation taxes our system and decreases performance and quality of life. [171]

Elevated cytokines signal increased inflammation,

which activates sympathetic response, which decreases HRV. All the Internet warriors always scream, "Show me the studies!" Well, here they are. There is no shortage of damning data on inflammation.

Inflammation makes you less resilient, both physically and mentally. It compromises immune function, lowers HRV, decreases mental processing abilities, and reduces emotional resiliency.

Continuing to break down that massive conclusion, depression is linked to elevated levels of inflammatory cytokines and abnormal immune function.

In the last line of that quote, they state that "depression is directly linked to decreased HRV," which we know decreases emotional resiliency.

How does acetylcholine fit into all of this? Well, recall that this study was investigating the cholinergic anti-inflammatory pathway. According to the study, "decreased vagus nerve activity, and the associated loss of the tonic inhibitory influence of the cholinergic anti-inflammatory pathway on innate immune responses and cytokine release, may enable significantly enhanced cytokine responses to stimuli that would have been otherwise harmless in the presence of a functioning neural circuit."

Under normal conditions, the cholinergic anti-inflammatory pathway exerts "tonic inhibitory influence on innate immune responses to infection and tissue injury."

In other words, vagus nerve signaling uses cholinergic pathways to regulate normal immune system function.

Since the research in still in its infancy, I cannot

give you a conclusive answer at this time about choline's role in immunity and HRV, but there are ongoing and planned future studies to examine the effectiveness and possible applications for this knowledge.

Here's what we know at this time: choline directly influences mental processing speed, including incoming and outgoing data – both visual and verbal. It also indirectly influences brain power through its role in immune system function via the cholinergic anti-inflammatory pathway and subsequent impacts that inflammation exerts on vagal tone/HRV.

GABA

GABA, short for gamma-Aminobutyric acid, is the inhibitory neurotransmitter, or the brakes that regulate neural excitability. [172, 173]

Adequate levels of GABA are associated with feelings of calm, relaxation, and social ease. Unlike the other neurotransmitters we'll discuss, GABA is responsible for slowing down our brains.

That may sound counterintuitive, but in today's world of non-stop connection, social media, twenty-four-hour news and sports, and on-demand entertainment that creates an overwhelming, sometimes debilitating sense of information overload, this neural balance is more important than ever.

Think of GABA as the governor for our neural system. It helps protect our bandwidth, so we don't cross that limbic system threshold.

Consider this study from 2007 that calculated the average person consumed enough information to fill 174 x 85-page newspapers per day. Eighty-five pages of information. Multiplied by 174. [174]

Think about that. If you sat on a deserted island in isolation and read three pages of a scientific study, your brain would be exhausted.

We're not living on deserted islands - we're living in chaotic concrete jungles with twenty-four-hour news, sports, and weather, and constant connection to our email, Facebook, Twitter, YouTube and more.

And we're not reading three science journal articles per day – we're reading 14,790 pages of information every day. No wonder so many of us are facing burnout and depression or turning to alcohol.

That study also found the same person now produces six of those eighty-five-page papers every single day through email, text, and hand-written communications.

By comparison, that same average person only produced 2.5 pages in 1986. We're still the same species we were 200, 2,000, even 20,000 years ago.

Our world and our lives are changing faster than our biology can adapt and evolve. It's up to us to realize this and manage our lifestyles in a way that positively impacts our health and well-being.

Keep in mind that study is ten years old. Ten years ago, we did not have #NetflixAndChill, Instagram or SnapChat. On-demand television, Facebook and Twitter were still in their infancy – far from the societal norms that they have become in 2017.

I'm comfortable predicting that our daily information consumption and production has increased by at least 50% since 2007.

This sensory bombardment and information

overload leads to something neuroscientists call "excitotoxicity:" the pathological process by which neurons are damaged and killed by the over-activation of receptors for the excitatory neurotransmitter glutamate.

Glutamate is the opposite of GABA. Its role is excitation, while GABA is charged with inhibiting or balancing this excitement. But we lead lives that constantly stimulate and excite our neurons.

I'd go so far as to explain excitotoxicity with the analogy of type II diabetes. In type 2 diabetes, our typical American lifestyle exhausts our insulin receptors over a lifetime of over-consumption of foods that elevate our blood sugar.

Between too little movement and too much unchecked ingestion of carbohydrate-rich foods, we force our pancreas to continually produce more and more insulin, as our cells become desensitized to its effects.

Over time, we find ourselves with cells that don't respond to insulin (enter pharmaceutical blood sugar control) and an overtaxed, dysfunctional pancreas (enter insulin injections, pumps, even pancreatic cancer).

Let's be clear: I'm not saying this is what will happen to our brains with excitotoxicity. Truth is, we don't know how our current lifestyle of constant sensory overload will impact our brains over time. We don't yet have the long-term studies on GABA and excitotoxicity.

But what we do know is it's not going to get better without mindful technology habits. Our ever-increasing information consumption isn't going to flatten out either. As our population increases and technology makes connection and sharing easier and faster, we're likely to continue to increase the amount of information we take in and put out on a daily basis.

This information gives weight to the importance of creating mindful technology habits, slowing down, finding balance, and periodically limiting sensory input.

Here's more evidence of our need to wind down and shut it out. The book *Stealing Fire* by Jamie Wheal and Steven Kotler details the $4 trillion "altered states" economy that reveals the magnitude of our desire to get outside our own heads. While some pursue these states for good, like Silicon Valley CEOs who use microdoses of psychedelics to solve global issues, others use it for distraction or numbing of the mind – like our TV, porn, gambling, and alcohol addictions that we use to escape our daily lives through altered states.

Since we're talking about the neurotransmitter GABA, let's focus on alcohol. According to a 2015 report from the NIH, 86% of the population over age 18 has consumed alcohol at least once in their life. 56% have consumed alcohol in the last month, and 27% have "engaged in binge drinking in the last month."

In 2010, alcohol misuse behaviors cost the United States $249.0 billion and "three-quarters of the total cost of alcohol misuse is related to binge drinking." [175]

If we know alcohol comes with such a heavy cost and can be detrimental to our long-term health, then why is alcohol such a popular tool for altered states or escape?

There are several factors that can't be ignored, including societal norms, but let's focus on the fact that alcohol is a GABA agonist. That means it binds to the GABA receptors in our brain, tricking our brains into thinking GABA is present. Consuming alcohol provides the same neural inhibition effect as GABA – with a side of inebriation, of course.

This explains why many go for a drink every night to "wind down" or before social events. We're chasing the "neural inhibition" as a solution for neural excitotoxicity more than we're chasing the inebriation. (Alcohol in social situations also meets our needs for another bliss-inducing neurotransmitter, oxytocin, which we'll discuss later).

Getting that GABA hit provides us with the neural inhibition that leads to the sought-after feelings of calm, relaxation, and social ease. And let's face it: we live in a world where most people "don't have time to be healthy" and instant gratification is king, so naturally grabbing a cold one or meeting the team for happy hour after work is easier, faster, and sometimes more socially accepted than seeking lifestyle habits and practices that promote long-term health.

This section is not about your personal choice to consume or not consume alcohol; it's about showing just how much our culture is seeking out this neural inhibition and identifying *why* you're using alcohol, if you are using it. Are you using it to escape or shut off your brain? If so, there are healthier alternatives.

Stealing Fire sheds light on the total cost of our pursuit of altered states, establishing our society's clear and alarming need for an answer to neural overload (low to moderate chronic excitotoxity).

I'm echoing the authors' plea that if we use altered states for good, and not to be distracted, numbed, or to escape our lives so we can get through without addressing what needs to be addressed, then purposeful and mindful shifts in both traits and states can be used to optimize and enhance our lives.

Remember: GABA is the only neurotransmitter that has the ability to slow the system down. It's up to us to realize that and provide it as much support as we can through our daily practices.

GABA is so powerful in this realm that supplementation has been shown to increase HRV, indicating that it promotes stress relief by helping us shift from sympathetic states to parasympathetic states. [176]

This doesn't necessarily mean you have to run out and buy a GABA supplement. There are plenty of parasympathetic state activities we can build into our lives to help support GABA and its role of neural inhibition. These healthy practices that help reduce neural excitation include yoga, meditation, breathe work, walks on the beach or hikes in the woods, float tanks that provide sensory deprivation, and neurofeedback.

A 2012 study looking into yoga's impact on the nervous system through vagal nerve stimulation found far-reaching implications from yoga's ability to shift the person from sympathetic to parasympathetic states (as measured by increased HRV), elevate GABA levels, and provide stress relief for subjects with depression, epilepsy, PTSD, and chronic pain. [177]

Anandamide and Oxytocin

Two other neurochemicals worth mentioning are oxytocin and anandamide.

Oxytocin is a neuropeptide hormone that plays a role in social bonding. Associated with companionship, it is released during childbirth, by hugging and singing, and it is involved in Stephen Porges's polyvagal theory. [178, 179, 180]

Porges does a lot of research on oxytocin. He asserts the saying "love heals" to be true to an extent, documenting in his research that "oxytocin can facilitate adult neurogenesis and tissue repair, especially after a stressful experience. We know that oxytocin has direct anti-inflammatory and anti-oxidant properties in in vitro models of atherosclerosis. The heart seems to rely on oxytocin as part of a normal process of protection and self-healing." [180, 181]

As Porges writes, oxytocin may actually help heal a broken heart. It exerts anti-inflammatory and antioxidants effects, promotes the growth of new nerve cells and tissue repair after stressful events, and appears to be part of the normal healing process. Porges is pursuing this research and its tremendous implication in PTSD treatments and socialization dysfunctions. For us, it's powerful evidence that communities, companionship, and support are crucial components of overall well-being, vagal tone, and cognitive function (decision-making).

Perhaps the most cited oxytocin research centers around the monogamous prairie voles who mate for life "because of oxytocin receptors while montane vole engage in short term relations because of lack of them." [182] Humans, another mammalian species with oxytocin receptors, have also been known to mate monogamously for life.

Another crucial neurochemical in the "flow cocktail," as Jamie Wheal and Steven Kotler describe it, is anandamide.

Anadamide is a neurotransmitter derived from arachadonic acid (AA), an omega-6 fatty acid that triggers muscle tissue repair and is commonly used by bodybuilders to "trick" the body into more muscle tissue repair. Dietary sources of AA include meat, dairy, and eggs. The word "anandamide" is derived from the Sanskrit word *ananda* for joy or bliss. [183, 184, 185, 186]

First discovered in 1992, anandamide is mediated by cannabinoid receptors in the nervous system and may have a role in immune function.

Thought to be connected to lateral thinking, anandamide has been shown to impair working memory in rats; it may also be responsible for the famous "runner's high" and is found in chocolate. [187, 188, 189]

These neurochemicals are part of an incredibly intricate relationship with the gut and brain and are inextricably linked to gut function, vagal tone, and cognitive function – especially as it relates to emotions, feelings, and mood. Zooming out one level, let's get back to our gut-brain connection and how research continues to link gut health to vagal tone/HRV.

Before we outline a few of my favorite tools

for gut-brain optimization, let's reflect on how our neurovisceral axis may be driving our decisions more than we realize and what we can do about it.

Reflection Questions

1. Are your dopamine needs being met naturally? Or are you engaging in dopamine-seeking behaviors?
2. What decisions are you making and what actions are you taking in an effort to seek safety and comfort?
3. How frequently are you engaging in "distraction" activities like TV, social media, eating, etc.? Ask yourself what you are really seeking when you engage in these behaviors.
4. How can you reduce, limit, or temporarily "detox" from incoming information?

Activities

Be in nature. Get Outside. Disconnect.

Or maybe this is better phrased as "reconnect with nature to balance your primitive biology." Our interaction with the natural world plays a crucial role in our overall health, perhaps nowhere as noticeably as our neurotransmitters.

We live in an artificial world. I'm not talking about social media (yet). I'm talking about our artificially lit, climate-controlled buildings and first-world problems.

There are exceptions, but as a whole, our society is soft and coddled, and it would be an embarrassment to our ancestors from 200 years ago – much less those who lived 2,000 years ago.

We evolved walking five to ten miles a day in search of food, water, and shelter. Now we freak out when the credit card swiper at 24/7 food mega-center is out of order or there is a line at the gas station where we pay $2-4/gallon so we can drive to buy food or go to work, where we sit all day with terrible posture and stare at screens that harm our eyes.

I'm not suggesting we revert back to the Stone Age. I'm simply pointing out how far (and how quickly) we have moved away from the life we biologically adapted to live. No wonder we're fat, sick, and unhappy.

It's up to us individually, then collectively, to

consciously employ and enjoy the fruits of our societal advancements and to do so mindfully, in a manner that enhances our health rather than destroying it. My intention with this book is to deliver you a multitude of methods to achieve this in your own life. Two such tools are natural light and time spent in nature.

Let's start with light. Like it or not, light has always played a massive role in our existence, and it transmits information to our biology. When our hormones and neurotransmitters are out of balance, our emotional resiliency and decision-making processes suffer.

Looking again to our evolutionary biology, we evolved waking and sleeping with the sun. As such, we have an internal biological clock that we also call our circadian rhythm.

In this natural rhythm, cortisol (a stress hormone) peaks in the morning to help us wake up, activate, and get moving. When functioning properly, it should bottom out at night as we prepare for sleep. [190]

An overlooked aspect of our ancestors waking with the sun is the exposure to morning sunlight. Our neurotransmitter production relies on this, as dopamine, serotonin, and eventually melatonin are regulated by morning sun exposure – specifically retinal exposure. [191]

That means no sunglasses and preferably no lens at all. As a contacts and glasses wearer, I get my morning sunlight sans glasses or contacts, so I know it can be done. [192]

This retinal exposure to the sun stimulates dopamine production, which fuels initiative and drive. It also stimulates serotonin production, the other feel-good neurotransmitter. It is this serotonin that is stored and then released after dark for melatonin production during sleep. [193]

Why is this important? The impact is twofold: first, a lack of morning sun exposure leads to maladaptive neurotransmitter levels and function. Second, melatonin production relies on this but is also disrupted by exposure to blue light after dark.

Blue light is a part of the visible spectrum of light that our brain associates with daylight. Morning and evening light have red and orange spectrum light, while midday appears more white or blue.

As part of our natural rhythm, when the sun goes down, there is no more blue light. Without interference from artificial blue light, in the two to four hours post sundown, we would begin to take the serotonin produced from morning sun exposure and convert it to melatonin. [194]

As eluded to above, we live in an artificial world, filled with blue light from artificial indoor lighting. The number one source of blue light, however, is from our screens – computers, phones, and TVs. This high exposure to blue light outside of our natural patterns disrupts melatonin production and sleep quality. [195, 196]

Americans now average more than ten hours a day in front of screens. Our eyes and our neurotransmitters are not adjusted to this. It's up to us to take precautionary steps to mitigate the negative impact of this new societal norm. [197]

Here's how. First, and for me this is non-negotiable, install blue-light-reducing software on your computer.

Flux is the most common, but a new technology from Belgian programmer Daniel Georgiev called Iris gives the user more control and has eye-saving features beyond blue blocking that I'll outline below.

You can also wear blue-blocker glasses, use screen protectors, and activate night shift mode on your smart phone to reduce screen brightness and blue light emissions after sun down. There is also Drift technology for your TV screen, but keep in mind that none of these is as effective as Iris at blocking 100% of the blue light emissions.

Some other tips I learned from Iris creator Daniel Georgiev include reducing the brightness on your computer, monitoring flicker rate which all LED screens have to prevent bulb burnout, using matte screens (like Kindles) whenever possible to reduce double images (that reflection of yourself you see in your screen significantly increases eye strain as your brain has to determine which image to focus on), and finally, take breaks from your screen every twenty minutes. Daniel recommends taking twenty seconds to look twenty feet away and blink a lot. He calls this his 20-20-20 rule. [198]

How to avoid or reduce exposure

Option 1 is to go supremely old school and romantic with all of your lighting coming from candles and fire rather than electricity. This will reduce brightness and optimize the ratio of blue to red light.

Option 2 is to swap out traditional light bulbs for red/orange spectrum bulbs that can now be acquired at any lighting store. Even Amazon has tons of options. Some people may not wish to swap their bulbs completely, so you can have certain light fixtures that you only use during the day and others that are loaded with red/orange bulbs for use after dark.

Quantum biologist and light expert Dr. Jack Kruse recommends using incandescent bulbs over LED or CFL bulbs if and when possible as they produce a more favorable ratio of blue:red light. [199]

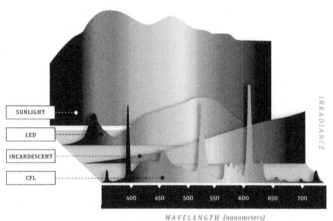

WAVELENGTH (nanometers)

Of course, limit screen usage after dark. If you must use a screen, use the tips mentioned above to optimize your usage.

Nature

As we shift our conversation from light to nature, let's briefly touch on cell phones, Wi-Fi, and non-native electromagnetic frequencies.

Cell phones and Wi-Fi have been linked to increased risks for brain cancer, and as little as twenty minutes of cell phone usage begins to disrupt cellular

DNA. [200, 201, 202] If you want a deeper dive on this, I highly recommend Katie Singer's *An Electronic Silent Spring*. It's full of legal precedent, scientific studies, and everything you need to know for a mindful, conscious, healthy use of technology.

Again, these technologies have exponentially advanced our civilization, so this is not about ditching the technology. Rather, it's about finding ways to limit personal exposure to protect our cellular function and improve health.

My favorite way to limit exposure and improve your health is to keep your phone on airplane mode when you're not using it. Next, is to avoid keeping it in your pocket. Leave it on a nearby table, desk, or in your bag. The greater the distance from your body, the better.

Turning Wi-Fi off at night can improve sleep quality and reduce daily (perhaps lifetime) exposure by 33%, based on eight hours without Wi-Fi out of every twenty-four-hour period. Or you can ditch the at-home Wi-Fi altogether and use Ethernet cables to reduce exposure even more. Remember: your neighbor's Wi-Fi counts too, as does that smart meter in your neighborhood, your Bluetooth headphones, and everything else emitting radiofrequencies around and through your body.

One of the best ways to balance in this area is to spend time in and expose ourselves to nature. You can do this almost anywhere, and as an added bonus, it's usually free.

Humans have spent 99% of our time on earth living in nature. Early civilizations lived in (and depended on) nature. Only since the industrial revolution and the hockey-stick-shaped growth curve of our population have we resided in urban areas.

Today, 50% of humanity lives in urban areas. This number is projected to grow to 70% in our lifetimes. [203] And as Singularity University Chairman Peter Diamandis predicts, 100% of the population (8 billion people) will be online by 2022-2025.

This isn't natural; it's not how we're wired or how we evolved. These urban living environments have their benefits, but they're also in direct opposition to our biology, and they're wreaking havoc on us. Our penchant to "get away" should be clue enough, but there are numerous scientific studies supporting this as well.

Take this conclusion from the same study that projects our increased urbanization, linking our removal from natural living to dysfunctional cognitive function: "urbanization is associated with increased levels of mental illness."

This same Stanford University study that found "city dwellers have a 20 percent higher risk of anxiety disorders and a 40 percent higher risk of mood disorders as compared to people in rural areas. People born and raised in cities are twice as likely to develop schizophrenia."

The science is quite compelling: more time in nature is linked to increased mental function and improved concentration – even in individuals with ADHD. [204]

Walking outside has been shown to increase creativity, with yet another Stanford study reporting that "walking opens up the free flow of ideas and it is a simple and robust solution to the goals of increasing creativity and increasing physical activity." [205]

Research from the University of Michigan has shown that walking in nature restores attention and improves memory functions in the brain better than walking in urban areas, where our brains must work harder to block out billboards, cars, horns, and other potential threats. In fact, the subjects enjoyed a 20% boost in memory after an hour in nature.

They also found in a follow-up study that we can get the same benefits by simply looking at pictures of nature. Pictures of urban landscapes did not produce the same memory-boosting effects. [206]

Not only does nature increase brain function, but it also boosts our mood and feelings of well-being. Stanford researchers have linked time in nature to reduced depression and stated it "may be vital for mental health in our rapidly urbanizing world." [207]

In a 2005 study, subjects who viewed nature images or spent comparatively more time in nature scored higher in self-esteem than subjects who viewed urban images. [208] A new 2017 study confirms the "existing evidence of the benefits of Natural Outdoor Environments (NOE) for people's health. It also suggest NOE potential as preventative medicine, specifically focusing on people with indications of psychological distress." [209]

This 2017 study is particularly interesting as the researchers looked at a value called Total Mood Disturbance (TMD) and quantified this data with HRV. This is further evidence to support our "everything is everything" theory.

Nature has also been linked to increased longevity, as one 2008 study correlated daily time outdoors with increased quality of life and life expectancy. Several other epidemiological studies have linked gardening to reduced stress and increased longevity in aging populations. [210, 211, 212]

> **Speaking of gardening and longevity, Dan Buettner's book** *The Blue Zones* **studied populations of people throughout the world who regularly live to 100 or more. One common thread among each of these populations, no matter their location, was regular outdoor activity – especially tending their own garden. Gardening can reduce stress, improve mindfulness, and as noted by Buettner, can expose you to a specific bacterium (mycobacterium vaccae) that could also benefit longevity. This bacterium has been shown to increase serotonin and boost mood as well as antidepressants. [213]**

I'm not saying you need to become a gardening aficionado, but gardening is a method of connection with nature, and if you live in a more urban area, gardening may be an option to increase your exposure to the benefits of nature.

We don't have to become hermits and move into the woods to mitigate these psychological and physiological impacts of civilization. We can reduce stress, improve mood, and reduce feelings of depression with short, twenty-minute walks in greens spaces with trees instead of highways. We can boost creativity, concentration, and increase longevity as well with equally easy-to-implement periods of time away from our screens and cubicles.

It can be as simple as taking daily walks outside. Find a green space, trail, path, or blaze your own. Just get outside and walk for twenty minutes per day. A 2012 study from the U.K. found that the color green boosts mood and makes exercise feel easier. [214]

Another way to increase your time outside is to implement walking meetings. Anyone who has ever worked in an office knows how brutal meetings can be, both for moral and for productivity. We also know that walking increases brain activity and that nature boosts creativity, reduces stress, and improves happiness. Leverage these benefits for a meeting where attendees are more engaged and productive and get a positive health return.

There is one downside (or potential upside, if you're looking to streamline meeting size): space is limited – walking meetings don't work for large groups.

Negative ions

Another benefit of connecting with nature involves our literal connection with the earth. Known as "grounding," or "earthing," it involves direct skin to earth contact. You can touch (or hug) a tree, walk barefoot on natural surfaces, get in a stream or river, or stand next to waves or waterfalls.

The earth has a negative charge, and this contact with the earth allows our bodies to absorb some of these extra electrons, giving us a negative charge of our own. [215]

A recent study on grounding showed that twenty minutes of contact with earth increases circulation throughout the whole body – not just the areas in direct contact with the earth. [216]

This increased circulation is good for reduced inflammation and improved cognitive function. Think of walking through the forest and coming across two different water structures. One is a cool, clear, running mountain stream, and the other is a stale, scummy, stagnant pond in the woods. Which one would you rather be in?

Float Tanks

I can't write this book and not mention float tanks. Also known as sensory deprivation tanks, these pods are designed to remove all sensory inputs for short sessions that will change your life.

Be warned: they're called sensory deprivation tanks for a reason. Not only will you be without your phone, you'll have your eyes closed, there will be no sound (unless you opt for music in some float spas), and you'll be floating in 10-14 inches of skin-temperature water saturated with close to 1,000 pounds of Epsom salts.

It's just you and your thoughts, floating in space. For an hour. Sounds like bliss for some, while it may sound excruciating for others. As with meditation, those who find this the toughest are usually those who need it most.

Personally, each of my float experiences has been different, yet each one seemed to deliver exactly what I needed at that moment. The uses and benefits are staggering.

Retired Navy SEALs are teaming up with concussion experts to use the tanks to help heal post-concussion issues, restore sleep, regain normal hormone function, and rewire neural pathways. [217]

Active-Duty Navy SEALs and other speed learners are also using these pods to reduce the time required to learn a foreign language from six months to six weeks. [218]

Athletes are using them to recover and heal faster due to reduced gravity and insane amounts of magnesium salts.

There aren't many studies yet on float tanks, but one optimistic study reported that "stress, depression, anxiety, and worst pain were significantly decreased whereas optimism and sleep quality significantly increased for the flotation-REST group. No significant results for the control group were seen. There was also a significant correlation between mindfulness in daily life and degree of altered states of consciousness during the relaxation in the flotation tank." [219]

As Float Seattle founder Sean McCormick told me on the OPP, high-pressure CEOs, entrepreneurs, and day-traders are using the tanks to silence the noise, reconnect and excel at their performance-based jobs. [220]

Everyday folks like you and I can use them to relax, recharge, explore consciousness, and reduce some of that excitotoxicity we've been talking about. Our GABA circuitry will thank us for this boost as we shut off all incoming signals for a short window of time. Find a float tank near you and enjoying some sensory deprivation. You can thank me later.

CHAPTER 6:
Brain Waves

That's the Yoga

Three words. That's all it took. A single three-word sentence uttered in less than two seconds back in 2014 changed my meditation and yoga practices forever. It's also changed my brain, my health, and my overall level of consciousness and awareness.

My wife (then girlfriend) Donna and I were on our first ever real vacation together in Orlando, Florida in August of 2014. Since we did not have access to a true Bikram Yoga studio at home, we decided to buy a week-long pass and attend as many classes as we could while visiting.

I had read about the benefits of Bikram for years and wanted desperately to experience it for myself. At that point, I had taken many yoga classes, including several hot yoga classes, but I'd never had the true Bikram experience.

For the record, hot yoga, no matter what they tell you, is *not* the same as Bikram Yoga. Bikram is a ninety-minute practice with twenty-six specific poses performed in a special sequence in a 100-108-degree room at 40% humidity.

As we left the studio following my first class, the instructor asked me about my experience.

My answer: "Loved it! It was everything I imagined it would be, but I found myself losing mental focus after forty-five minutes, and I struggled to stay present, especially the last thirty minutes."

That's when he dropped the bomb on me: "That's the yoga."

Simple. Profound. And, as the years go by, I'm so grateful that my first experience was with a practitioner who could impart so much profound wisdom with so few words.

To this day, I use this mantra in my meditations, in yoga practice, and even when I do cold exposure.

At its essence, his response said: if you can learn to control your mind, you can do anything.

Let's explore why this is true from a neuroscientist's perspective.

Brain Waves

"You're not going to stab my skull with that, are you?" I asked hesitantly.

"No!" was the laughter-filled reply.

Jessica, my lab technician for the day, was used to this question.

The syringe in her hand, while daunting in size, was merely plastic and was only being called upon to insert gel under the twenty-eight electrodes held to my head by a cloth skull cap that resembled Michael Phelps's swim cap. This contraption on my head was a wet EEG machine (electroencephalogram), being used to map my brain waves.

As we outlined previously, neurotransmitters are the chemical messengers that carry communications through our nervous system. This collective synchronization of electrical pulses of communications between neurons is referred to as "brain waves."

If neurotransmitters are the language of neural communication, brain waves are the sound of the crowd, or the noise profile of the room. Brain waves literally control every aspect of measurable brain function, from focus and memory to executive function. Like the palpable atmosphere in any arena, brain waves determine our mood, emotion, and behaviors.

I'm at Peak Brain Institute in Los Angeles, California with neuroscientist Dr. Andrew Hill and the EGG cap on my head is being employed to get a baseline reading for my brain waves.

Peak Brain uses something called QEEG (Quantitative Electroencephalogram), where our brain waves are compared to a database of brain waves collected from people of the same age, gender, and health.

The scan detects any previous brain injury or trauma (I had none) and reports on the current (baseline) environment of brain waves in the wearer's head.

The first of my two readings is done early in the morning, fasted, and without any caffeine, vitamins, or medications, in an effort to keep the reading as pure as possible. This gives the best glimpse possible of what is going on inside our brains.

There are five types of brain waves neuroscientists measure: alpha, beta, gamma, delta, and theta brain waves.

Alpha brain waves were the first to be discovered by German neurologist Hans Berger who invented the EEG for brain-wave mapping, hence their name.

What's interesting about alpha brain waves is that the frequency of these waves (7.5 - 12 Hz) very closely resembles the resonate frequency of the earth (7.83), as discovered by German physicist Winfried Otto Shumann, for whom the Schumann frequency is named. [221, 222, 223]

Once again, in our "everything is everything" theory, the fact that our "default" brain-wave status that matches meditation and mindfulness also matches the vibrational frequency of the earth is pretty profound.

Beta brain waves are present when we are alert, attentive, engaged in problem-solving, judgment, decision-making, and focused mental activity. These waves are characterized by low amplitude, are the fastest moving, and correspond to the highest level of brain activity. The frequency of beta waves ranges from 13 to 30 Hz.

Lower in frequency and higher in amplitude, alpha brain waves reflect non-arousal. These are not sleep states, however. Instead, these waves are seen in relaxed states of wakefulness like meditation or contemplative states. Alpha-brain-wave frequency ranges from 7.5 to 12 Hz.

Continuing to increase amplitude and reduce speed, we have theta brain waves. These brain waves are associated with flow states. Runners or other exercisers who get lost in their thoughts and come up with numerous great ideas can attribute those breakthroughs to theta states. Theta-brain-wave frequency ranges between 4 and 8 Hz.

Delta brain waves are the deepest, slowest waves, and they're associated with sleep. Delta frequencies range from 1 to 4 Hz, with deep dreamless sleep being the lowest frequency and lowest brain activity. (They never hit zero, as that would correspond to zero brain activity or being brain-dead).

At 30-70Hz, gamma brain waves are by far the fastest, and their full impact is still being researched. Some researchers suggest these fast brain waves are optimal for high brain activity and states of learning. Some researchers believe these brain waves are associated with the unification of conscious perception, but others disagree, arguing that this cannot be stated with certainty at this time.

What we do know is that we can entrain specific brain-wave patterns to elicit desirable mental states for elevated performance.

Optimizing Brain Waves for Performance

Back in the 1970s, brain waves were being utilized to mold performance by Ned Herrmann, the leader of management education at General Electric, where he integrated brain activity into his management training principles.

Ned summarized the importance of brain-wave awareness in his work here: "It has been my personal experience that knowledge of brainwave states enhances a person's ability to make use of the specialized characteristics of those states: these include being mentally productive across a wide range of activities, such as being intensely focused, relaxed, creative and in restful sleep."

As I learned from Dr. Hill, it's important to be aware that while one brain-wave pattern may dominate, we always have some of each brain wave present, in varying levels. So, for sustainable high performance, the question becomes: what is the predominant wave pattern at any given time and what do the ratios look like?

Some ratios can have seemingly disadvantaged outcomes. For example, a high alpha:beta ratio correlates with behaviors like inattention (ADD), being stuck in neutral, and being a big-picture person with fast-moving thoughts.

Conversely, a high alpha:theta ratio corresponds to behaviors that display impulsiveness (ADHD), being an action-taker, and being a quick, "live in the moment" type of person.

I'd also like to make my clinical friends happy and insert this here for clarity and accuracy: per the American Psychiatric Association's Diagnostic and Statistical Manual of Mental Disorders (DSM) 5, there is no longer ADD, only ADHD with "types," i.e., Inattentive, Hyperactive or Combined Type.

From an objective viewpoint, none of those traits is inherently positive or negative. They are, like any tool however, better or worse for certain applications.

Dr. Hill agrees, reminding us that "ADD is only pathology if we force you to sit in a chair without moving for 8-10 hours."

Brain-wave awareness is about understanding the strengths and weaknesses of each state and being able to use them to our advantage while training our brains over the long term to produce a most favorable state for our given pursuits.

Wondering about my results? We took two readings, the first on a Wednesday morning as a baseline and the second on a Friday morning to test Natural Stack's Nootropic Stack CILTEP, a supplement designed to boost focus and memory.

My baseline QEEG showed a clean brain with no trauma or injuries. However, my brain-wave pattern reflected the aforementioned high alpha:beta ratio – congruent with my ADD tendencies.

My alpha brain waves were 2-3 standard deviations higher than the norm. This objective reading validated my years of dealing with inattention, feeling stuck in neutral, seeing the big picture, and thinking about huge, massive undertakings while struggling with the day-to-day monotony.

This is what prompted Dr. Hill's statement: "ADD is only pathology if we force you to sit in a chair without moving for 8-10 hours."

It's also one of the reasons I've had to find systems that keep me on track and moving forward, like the "Move the Chains" analogy.

All strengths come from what were once deficits, and this is certainly one of the reasons why systems and processes have become such a part of my journey.

Armed with the knowledge of how my brain works and where it thrives, I can tailor my tasks to best suit those characteristics.

In my second reading, a single dose of CILTEP lowered my alpha brain waves by 1-2 standard deviations, bringing my brain instantly into a more focused state.

While encouraging for CILTEP as a tool in our toolbox, it's important to point out that this was an experiment of $n=1$, as we like to say in the quantified self-world. Natural Stacks is in the process of testing on a larger scale, so stay tuned for the results of those clinical trials.

To me, this finding is the real key of all self-quantification pursuits – be it QEEG, wearable fitness trackers, or blood ketones monitoring – to be able to match objective data with subjective feelings so we can learn what each *feels* like, so you can eventually recognize it without having to always rely on tools or technology.

As the saying goes, "He who masters himself has true power." The technology at our disposal today can dramatically shorten the learning curve to self-mastery if we use it to help us in that pursuit.

So what do we do with this data? How can we use it to help us regulate feelings and emotions and drive high performance?

QEEG is about establishing a baseline so neuroscientists know which direction is the best course of action forward. At Peak Brain Institute, the follow-up to this QEEG is neurofeedback brain training.

Neurofeedback and binaural beats are two increasingly popular methods of optimizing brain waves for peak performance. Both work through a process called "entrainment." Entrainment is the phenomenon through which our internal brain waves sync to and match external wave inputs.

Let's talk about neurofeedback first because it was what we used following my QEEG readings and because I think it is a superior long-term strategy than binaural beats. I see neurofeedback as a practice that can impact both short-term states and long-term traits, while I'm not sure that binaural beats are anything more than state enhancers.

Neurofeedback is a method of direct training of brain function where biofeedback is sent directly to the brain, so it learns to function more efficiently. The term "biofeedback" implies that a yes or no signal is given directly to the brain.

Essentially, it is a slap on the wrist for undesirable brain wave patterns or rewarding the brain for changing its own activity to more appropriate patterns. It's important to note that these changes are very gradual, but they do change the brain faster than meditation (two or three months versus a lifetime of practice).

With neurofeedback, a device is worn on the head to connect electrodes to the skull. These electrodes emit wave patterns that correspond to the desired brain waves, essentially telling our brains what environment to create.

Most neurofeedback interfaces have a screen with a flying spaceship or a ball that the user appears to be controlling. Despite its game-like appearance, neurofeedback is not an actively controlled training program. In other words, the spaceship on the screen is not something you consciously control. Instead, it is more like a pat on the back that rewards desired behavior. If the brain shifts its frequencies away from the desired pattern, the spaceship crashes, serving as a slap on the wrist of sorts to teach the brain that those waves were not the desired waves. When the brain's waves match the desired scenario, the spaceship flies, hence the biofeedback and reward loop.

By starting with a baseline QEEG reading, a neurofeedback practitioner is able to outline a program for two to three months of neurofeedback training that can optimize any aspect of brain function that we can measure.

The baseline testing is crucial, as we all have unique brainwave patterns. QEEG serves as a "you are here" sticker on the map to optimal brain function. Without this starting point, the use of neurofeedback is a "spray and pray" approach. It's nearly impossible

– certainly unadvisable – to chart a path without first knowing your starting point.

The future of this field is exciting, as with the right practitioner, we can train *any* brain dysregulation from brain waves, to healing traumatic brain injuries, to improving sleep, and more.

In my case, this training program was designed to decrease alpha brain waves and slightly boost beta brain waves, reducing my inattention.

The take-home with brain waves is the same as with any other path to growth: find gaps and create programs that train your brain to produce a more favorable environment for you and your mission. One more benefit of neurofeedback is the long-term protection it offers against cognitive decline.

As he explained when I interviewed him, Dr. Andrew Hill points out that "age-related cognitive impairment is driven by loss of cortex - cell bodies in lateral parts of frontal lobe and the incileal cortex (temples) - that correspond to body awareness, feeding, food appreciation, self-control." [224]

Just how prevalent is age-related cognitive decline? According to Dr. Hill, the average person experiences a 10-25% structural loss after the age of sixty-five years old.

In the research, only one demographic was spared this structural brain loss – those who are lifetime meditators. This is further evidence that meditation not only helps us create short-term shifts from sympathetic to parasympathetic, moving from limbic system control to prefrontal cortex control (thus enabling higher-level consciousness), but also has long-term, quality of life benefits.

The research also found that neurofeedback training can deliver the same brain protection as a lifetime of meditation in as little as twelve weeks. This is the information that turned me on to neurofeedback and set in motion my trip to Peak Brain to meet with Dr. Hill.

My favorite part of neurofeedback training is that you can do it while you're doing your other activities. You don't need to dedicate time to *only* doing this, so it's easy to fit it into your day. You can do it while answering emails, cooking dinner, or driving to work. It's a non-invasive practice that trains your brain to produce favorable states and traits for both short-term benefit and long-term health. And it is this non-invasive use that increases compliance and consistency – something that we all know makes or breaks results.

The downside to neurofeedback is the price. At $5-$40,000 depending on the option you choose and practitioner, it's cost-prohibitive for many.

Binaural beats, on the other hand, are a much more cost-friendly option. They're free on YouTube, Pandora, and Spotify, but I suggest a paid service like Brain.fm for a high-quality product designed by scientists.

Brain.fm is backed by neuroscience and more than a decade of music creation. When interviewed for the podcast, they described their service as "specially designed music that improves focus, relaxation and sleep within ten minutes."

They also reminded us that "the creator of binaural beats said they don't work, but nobody actually reads the study!" (Adam Hewett, Co-Founder of Brain.fm).

Hewett is talking about biophysicist Gerald Oster, who in 1973 presented the original thesis for binaural beats. In that paper "Auditory Beats in The Brain," Oster examined the phenomenon of playing two tones (up to 26-30Hz apart), one in each ear, and how the brain perceived these as a "binaural beat." As Hewett reminds us, Oster never mentions brain waves or entrainment in this original paper. [225]

Binaural beats, as we know them today, are similar to neurofeedback in that they attempt to use entrainment to alter our brain waves. The term "binaural" refers to the fact that two different wave patterns are administered through headphones. The brain receives different wavelengths in the right and left ears, then syncs them up internally to create a single brain wave in the middle. By definition, binaural beats cannot be effective with only one speaker.

For the record, Hewett claims Brain.fm's music, designed to increase focus, relaxation, and sleep, picks up where traditional binaural beats leave off, inducing the desired changes in brain-wave profiles. Through entrainment, our own brain waves begin to mimic this input signal.

One problem with binaural beats is that any listener can choose any random "song" or brain-wave pattern at any time.

For an experienced user who recognizes their current state and knows what brain waves they need for the current state or task, this is not a problem. In fact, this choice is a benefit. But for beginners or intermediate users without intricate knowledge of their brain wave patterns, the associated moods and feelings, and what brain waves are best suited for the task at hand, this is akin to our neurofeedback use without a map or known starting point.

In these circumstances, the user could unwittingly move their brain *away* from the desired state. My recommendation, if you're going to use binaural beats, is to use Brain.fm.

Neurofeedback, unlike binaural beats, gives us the power change the traits – not just the temporary state – of our brains for characteristics and traits more favorable to peak performance.

Neurofeedback, when administered by a professional, is customized to your goals, and each session serves a specific purpose. I think of neurofeedback as a customized personal-training program for our brains, while binaural beats is like grabbing the coolest workout you found on social media for that day. Both are workouts that can serve a purpose on a given day, but one has greater long-term benefit.

Reflection Questions

- How well do you control your mental chatter?
- What is the environment of the arena within your head?
- What are you doing to direct it favorably? What habits and practices can you add to make sure this environment is favorable?
- Is meditation/yoga a part of your toolbox for optimization and self-mastery?

Activities

Music

Music has a profound impact on our emotional states. I know it, and you know it. But since we're talking about neuroscience, it's not enough to just say, "It works because we've both felt it."

Personally, music has always been a huge part of my life. I have my parents to thank for my good (and eclectic) taste in music. Before I could pronounce the letter "r" properly, I was in the back seat singing Peter Gabriel's *Sledgehammer* and begging "Mommy, mommy, can we listen Bwuce?" (as in Springsteen, The Boss of course, who to this day remains my all-time favorite musician). At least, this is how my aunt tells the story.

Absolutely nothing alters my mood as quickly or as powerfully as music. I'll never have the opportunity to experience this as another person, so it's forever an n=1 experience, but the science seems to support that this effect is species-wide, not just an individual experience.

Stephen Porges is a huge fan of music for elevating our states and has given talks detailing the effectiveness of ancient practices of group chanting, prayer, and songs on increasing vagal tone. [82]

<u>Music on brain activity</u>

Our brain's reaction to music is visible. Using functional MRI, Wake Forest neuroradiologist Jonathan Burdette watched the areas of the brain associated with thought, empathy, and self-awareness while subjects listened to music they liked and disliked across five different genres (country, rap, rock, classical, and Chinese opera).

He discovered that the music we like, regardless of the type of music, had the greatest impact on brain connectivity. Burdette calls this circuit that includes thought, empathy, and self-awareness the "default mode network," and found that it was poorly connected when exposed to disliked music but "better connected when listening to the music they liked and the most connected when listening to their favorites." [226]

This research is all the more reason to crank up your favorite tunes when you need an emotional boost or reboot. Just be careful not to drown your sorrows in sad songs if you're trying to elevate your mood.

A 2009 study from London, England shows that music makes our emotions stronger, whatever they are. For a musical analogy, it acts as an amplifier.

They found that "happy music made happy faces seem even happier while sad music exaggerated the melancholy of a frown. A similar effect was also observed with neutral faces. The simple moral is that the emotions of music are "cross-modal," and can easily spread from one sensory system to another." [227]

If you want to feel and explore your sadness, that's perfectly fine – just be aware that musical selection will take you further down that pathway. Conversely, if you're looking for a pick-me-up, choose music that makes you smile, laugh, or dance to move your brain toward being happy by first altering your physiology.

Smiling activates neurons in the brain associated with happiness. Studies show that people with depression symptoms experience less symptoms after forcing a full smile, even though nothing in life changed. One of the fastest ways to get to a genuine, full smile is to go all out as you dance to Footloose, MJ, or Prince. [228]

Music doesn't just impact our brain. It has a profound impact on water, something that comprises 2/3-3/4 of our being. Japanese researcher Masaru Emoto studies the crystal formations in water. He and his team freeze water and then photograph the crystals formed by different sources of water and analyze them under a microscope. The next step is to then expose water to different words, pictures, and music to see how those environmental stimuli affect the formation of crystals in the water.

Emoto and his team demonstrated that water collected from "holy sites," famous springs, and other healing waters around the world do in fact form nearly flawless crystals, while most tap or "regular" waters form zero or dysfunctional crystals. [229]

A note on these "healing waters." They're not crystalizing because their source is heralded by folklore; it's the other way around. Our ancestors, who were much more in tune with nature and with how the natural world impacted our bodies, discovered certain improvements in health, energy, and well-being when they drank from those sources. They, in turn, began to celebrate the water from those sources, and the legends were passed down. The fact that modern science can now definitively show a structural difference in the water from those sources is quite interesting.

What does this have to do with music? That's the second phase of the Emoto's experiments.

Positive words like thank you, love, peace, and gratitude helped even plain water form more perfect crystals. The same for classically composed music.

How does this happen? Emoto explains that music is sound, and sound is vibration. The unique vibrational patterns of musical compositions impact water differently. Music that vibrates at higher levels leads to improved crystal formation. Conversely, music and words (like fear, hate, ignorance) that vibrate at lower levels impaired crystal formation.

Since humans are mostly water, it stands to reason that words and music have the same effect on the water we drink, the food we eat, and the water inside our bodies. Granted, this would be tougher to study in humans, and I'm not volunteering to be frozen and have my liquid contents examined for crystalline formations.

To recap, positive, loving, grateful, and high-vibrational-status words and music resulted in crystal formations, while negative, lower vibrational words and music impaired crystal formation.

Classical music had the best impact on crystal structure, and the two words with the biggest impact were "love" and "gratitude." Negative words and phrases had negative impacts on the crystal structure, showing dysfunction and maladaptation.

This is powerful scientific evidence to support the more "woo-woo" beliefs in the power of vibration and positivity. Since the crystal formations are visual, I encourage you to check out his work, so you can see for yourself. For more on energy and vibrations, I also recommend the book *Power Vs. Force* by Dr. David Hawkins.

Here is a chart he provides in that book to correlate vibration equivalents for common words and feelings. [230]

Level	Emotion	Log
Enlightenment, various levels	Ineffable	700 – 1000
Peace	Bliss	600
Joy	Serenity	540
Love (unconditional)	Reverence	500
Reason	Understanding	400
Acceptance	Forgiveness	350
Willingness	Optimism	310
Neutrality	Trust	250
Courage	Affirmation	200
Pride	Scorn	175
Anger	Hate	150
Desire	Craving	125
Fear	Anxiety	100
Grief	Regret	75
Apathy	Despair	50
Guilt	Blame	30
Shame	Humiliation	20

As you can see, 200 is the line of demarcation. Anything below that level is "negative," and anything above it is "positive." Hawkins suggests (and Emoto's research supports) that simply looking at or repeating words at a higher vibrational status can help us move to higher levels of vibrations.

Here's to good vibes, vibrating higher, and

elevating consciousness.

Music also impacts HRV, but it's far from overwhelmingly positive. A 2012 study monitored HRV after subjects listened to both "relaxant baroque and excitatory heavy metal music." Contrary to what we might think, *both* groups saw decreased vagal tone. The researchers say these results may be due to volume. "We suggest that relaxant baroque and excitatory heavy metal music slightly decrease global heart rate variability because of the equivalent sound level." [231]

An earlier study in 2005 study also found that "excitative music decreased the activation of the parasympathetic nervous system." [232] Does this mean music is bad? No.

Recall that Porges himself lauds ancestral chanting, prayer, and singing as activities that increases vagal tone. Recall that each of the seven major chakras is associated with a musical note (pure vibrations). And don't forget about Masaru Emoto's crystallization of water through exposure to classical music. There is zero doubt in my mind that music can have a profound and positive impact on us.

This HRV research does suggest that excitatory music used to transiently elevate mood may be best reserved for short-term "state-hacking," as the temporary elevation in alertness and emotion is accompanied by decrease in HRV.

I can speak from experience: when I spent all day in my performance-training facility, the House of Strength, listening to "excitatory music" as the researchers call it, it sapped my adrenals. I can recall many days driving home in silence. Yes, there were other factors at play, music was not the only variable, but I've spoken with other gym owners who have similar experiences from being exposed to high-intensity music for long hours on a consistent basis.

Don't get me wrong; I love music for altering brain waves and physiology in the moment, boosting mood, and elevating our states, but let's be sure we realize it is a short-term fix and not one that we should rely on daily. Note the difference between "relying on" and "enjoying."

I can put on Bruce Springsteen and instantly feel calm and relaxed, yet hungry and motivated to create something more with this life. I can put on Michael Jackson or Prince Pandora stations and dance like a crazy person with no cares in the world. Or I can listen to Van Morrison, Led Zeppelin, or Mozart stations and be as chilled as a cucumber. I'm not going to list all my Pandora stations or Spotify playlists here; you get the point. Use different music to help balance or dial up the mood you need or want for any given day.

No two days are the same, and sometimes, it's a process to find the right one. Change it up, try different options, but one caution: don't spend all your focus picking the right music. If you try three or four types one day and it's not happening, move on to a different "hack." Don't lose sight of the fact that the music is there to enhance your work – you're getting paid to DJ your own day.

Yoga and Meditation

While we can utilize music as a transient, state-shifting tool in our toolbox, yoga and meditation have the power to positively alter our immediate state as well as our long-term traits.

Numerous studies have shown that yogic practices have tremendous benefits for mood, brain function, longevity, and more. Yoga works for all the same reasons that movement does, because it *is* movement, but it's also an intentional movement usually paired with breathing and/or meditative elements that provide extra benefits.

Those benefits include increased time in the parasympathetic state, increased HRV, and increased awareness. [233-237]

In a 2007 study from the University of Wisconsin-Madison, researchers found that yoga and meditation's ability to increase awareness can also help fine-tune control over our attention.

They repeated a famous study that investigated a phenomenon known as "attentional blink." In this study, subjects were shown multiple images in rapid succession, hidden among numbers. The closer together the images were placed, the greater the likelihood of missing or not recognizing the images. This "attentional blink" is caused by depletion of the neurotransmitter acetylcholine.

In the 2007 repeat of this study, subjects practiced meditation for three months and found that "three months of intensive meditation reduced brain-resource allocation to the first target, enabling practitioners to more often detect the second target with no compromise in their ability to detect the first target. These findings demonstrate that meditative training can improve performance on a novel task that requires trained attentional abilities." [233]

Aside from increased awareness and focus, a 2016 review of fifty-nine studies suggests that yoga increases vagal tone and HRV. The researchers do point out shortcomings in some of the studies and suggest that research continues to investigate. But fifty-nine studies with positive correlations is pretty strong evidence in favor of yoga as a parasympathetic-inducing, HRV boosting practice. [234]

Yoga and meditation are parasympathetic activities, turning off sympathetic fight or flight, shutting down defense responses, and as such, they support increased vagal tone and increase HRV – which, as we know, increases emotional resiliency. [235, 236, 237]

Both meditation and yoga involve focusing on the breath, a practice that helps consciously control the autonomic nervous system via the vagus nerve, which both controls and is influenced by breathing – again increasing presence, awareness, calmness, and vagal tone.

This increased awareness and connection to our center/chakras/true self through breathing affords us the opportunity to train and strengthen parasympathetic, or prefrontal cortex control over sympathetic, limbic activation.

Think of this relationship like walking a dog. We've all seen the overexcited dog that "walks the human" rather than the other way around – yanking, pulling, and otherwise running the show. Training and strengthening our prefrontal cortex to be the default pathway instead of letting the limbic system run the show is like taking that compulsive dog to obedience school. Now that dog heels and walks by your left foot without leash because it has learned that you, not it, are in control.

How to implement

I suggest finding the type of yoga and/or meditation and the frequency of practice that works best for you. Research suggests at least one yoga session per week and daily meditation, even if it's just for five minutes.

I'd also like to add that tai chi and chi gong are other practices shown to improve HRV through their activation of the parasympathetic nervous system. As such, these could be used over yoga, if you so desire. I do my own version of Chi gong on days that I do not lift weights to facilitate physical recovery and increase daily movement. [238]

Do more if you need or want more – how it fits into your life and your movement/training regimen requires more information than I have. If you need help implementing this or any of the tools outlined in this book, go to ryanmunsey.com/apply for more information on how to get coaching, consulting, or any of the massive amounts of free content I'm putting out on the podcast, blog, and social media.

Harris, it's worth noting, has a product called Holosync. This "meditation on steroids," as he calls it, is music that combines chakra-aligning crystal bowls with brain-wave-optimizing binaural beats designed to help increase awareness by training our brain to prioritize the prefrontal cortex over the limbic system.

Consider things like Holosync, neurofeedback,

or Brain.fm as tools to enhance your practices. You can use them with your meditation practice or yoga if doing it at home, plus they can be used as an "ace up your sleeve" for state-hacking to alter brain waves.

Make no mistake about it: yoga and meditation are a lifetime pursuit, as is mental health in general. There is no quick fix, and we're never "there." We must always remain vigilant over our minds and constantly chaperone our default beliefs and operating systems.

But again, it's a worthwhile pursuit – in the short term, it provides us better moods, thoughts, and feelings. In the intermediate, we get better days, health, choices, and relationships. In the long term, we enjoy reduced risk of cognitive decline and the fruits of a life well-lived (friendships, experiences, communities, and health).

Remember, "that's the yoga," grasshopper. That, in itself, is the practice.

Short-Term, Transient "State Hacks"

Brain.fm is my preferred method for binaural beats. Co-founder Adam Hewitt points out that Brain.fm is "music designed to improve focus, relaxation, and sleep," but it is not binaural beats. Recall that binaural beats don't actually succeed in their theorized goal of entraining brain waves. [239, 225]

Neurofeedback is great if you have access to it. However, it's not something you "feel" like a cup of coffee, so it may not be the best option to quickly alter your state if you need a jolt on any given day. I recommend keeping neurofeedback in your trait-development column.

As for yoga and meditation, they are definitely more long-term, trait-development practices, but they can have powerful state-altering effects for any given day as well. I can honestly say I've never regretted going to a class. Every time I go to Bikram Yoga, I leave renewed and refreshed – almost as if I had been baptized.

Pardon the biblical analogy, but it's the only way I can describe the spiritual, physical, and mental state cleansing and renewal that I feel after a Bikram class. I want to eat only the purest, cleanest foods, and I want to engage in only the cleanest, highest-vibrating thoughts. The typical bullshit from our modern world that normally clouds our minds is gone, and my being wants to reject its reentrance post-class. Yes, I hear myself, and I admit this sounds "woo-woo," but it's how I feel after these classes, and since this book is about optimizing feelings, thoughts, and the decisions they drive, this could not be more relevant.

Simply put, Bikram puts me in a monk-like trance where my emotions, feelings, and thoughts are clean, pure, and as enlightened as I've experienced. It's a place from which I'd like to operate on a daily basis. That is why I so strongly encourage you to find that thing for yourself. It may not be Bikram – and it may not be yoga at all – that's why this book is full of tools to add to your toolbox.

Experiment with all of these trait-developing and state-changing tools to find what works best for you and realize that some will work better than others in particular situations. Psychology pioneer Abraham Maslow said it best: "Everything looks like a nail, when you only have a hammer." Don't be a one-tool human.

Meditation

I know part of writing a book is providing a prescription for how to implement certain strategies. Meditation is the practice of focusing on and understanding one's own mind.

As such, I can only explain to you the benefits and the systems you're exploring. How you choose to explore this space and what understanding you gain from that search is up to you.

Meditation comes in many forms, and it's up to us as individuals to find the modalities we prefer. I've always found some of my best Headspace while driving, mowing grass, hiking, or cooking. I think there is something about the monotony of those activities that simultaneously keeps me engaged, yet lets my mind wander.

Meditation can be done in silence, with your eyes open or closed, alone or with a guide, with an app, focusing on breathing, or none of the above.

In the words of the most famous Zen philosopher from the Western world Alan Watts, "You can make almost any human activity into meditation simply by being completely with it and doing it just to do it."

If you're one of the millions who have self-professed struggles with meditation or mindfulness, stop wondering if you're doing it right and just do it. Stop looking for the result and focus on the act of *doing*. Remember: focus on actions, not outcomes.

If you want to try some tools or apps to jumpstart your meditation practice, some of the most popular ones include Calm, Headspace, Holosync, and Brain.fm.

You can also focus on breathing to give an active mind something to latch onto in an effort to begin to slow and quiet it. Try the breathing techniques included at the end of Chapter 7.

CHAPTER 7:
Mindset – Upgrade Your Operating System

Dream (and Fail) Like a Kid

"I'm going to be a fighter pilot!"

That's what the adorable young boy with blond hair and a toy jet proudly stated to the woman across from him in the airport terminal as we waited for our flight.

I couldn't look away. The conviction, the passion, the *certainty* in his voice were infectious.

Have you ever heard a child talk about what they want to do when they grow up?

Whether that vow is to fly fighter jets, be lawyers or doctors, fight fires, or become President of the United States, their passion and conviction is unrivaled. You can feel the certainty in their words and their beliefs. Even more interesting, is their complete disregard for the objections adults would feel.

Can you imagine an adult stating such a bold dream with that level of conviction? Most wouldn't. Most of us won't even allow ourselves to think it. As soon as the idea pops into our head, we're already crushing our own dreams with thoughts of how we could never do it, why we shouldn't try, and what others might think of us for talking about it, much less going for it.

If (and that's a big *if*) we allowed ourselves to utter those words aloud in public, we would do so only after considering how others would respond, what they would think, and crafting the perfect "sales pitch" with which to convince them and protect our egos.

We may even lower our voice when we speak it to reduce the chance that someone will overhear us. The strangest part of that action is that most the people who might overhear us are people we don't even know; why the hell do we care what *they* think? The answer lies in our lizard brain's hardwiring to please the tribe, avoid being different, and keeping our vagal tone high by feeling safe within the pales of societal norms.

Have you noticed that? Have you noticed how adults are "trained" to be average and discouraged from thinking big, dreaming, and sharing those dreams? But kids do none of that filtering. Their operating system is still pure. No matter our age, we all still have this ability.

Unlike (most) adults, kids are locked in on their dreams. Damn the details and damn what others think. What happens to us as we become adults?

Sure, we change as we grow and mature, and we may lose interest in the callings that our pre-pubescent selves proclaimed for us. And sure, every pursuit needs proper planning and risk management, but *something else changes* from childhood to adulthood.

It's not the callings themselves that change. It's us. We lose that passion, conviction, and complete lack of doubt, fear, self-sabotage, or negative self-talk that fades away.

Most of us have probably heard a child speak of these dreams, whether our own kids or kids at airports or other public places. They're sharing these dreams to anyone who'll listen. They have no shame. And that's part of the point.

While these experiences with children are

common – even universal – similar experiences with adults are significantly less common.

When was the last time you heard an adult speak of their dreams with such unfiltered passion, conviction, and complete lack of shame about you or anyone else might think?

It's a rare thing. But when we do encounter it, it's striking. It's infectious and inspiring.

The adult with child-like passion, conviction, and commitment stands out. They're committed, devoted, and utterly one with their passion.

I hope you have been fortunate enough to encounter a few of these lucky souls. The explorer, the entrepreneur, the blogger – the person who exited the matrix-like rat race, ignored the logical reasons not to, and stopped caring what outsiders think. The person who is intensely passionate and happy in their own skin each and every day, pursuing their true calling.

Perhaps these encounters inspire you, or at least make you think, "I wish I had that life," or, "I wish I was that passionate about something."

You are. We all are. And we can have that life if we get out of our own head and our own way and start moving towards our calling.

Forget this "finding your passion" stuff. We have to develop our passions. We all know what makes us light up. From there, it's a simple path: time, repetitions, and deep knowledge all build confidence and lead to mastery of our passions.

But we must take action. We must start, and we must persevere.

Sadly, most will never do this. We care too much

about what others might think. We worry about what they'll think of us if we try. And what they'll think of us if we fail. We worry about what they'll think of us if we actually succeed, leaving our past lives behind.

Do you think failure ever crosses the mind of a child when they think about or speak of their ambitions? Does any of this cross the mind of a child?

We overthink it. We overanalyze it. And then we start to believe the stories we tell ourselves about how and why it shouldn't happen for us. Remember: imagination creates a vision of our future. F*ck those thoughts and feelings!

If any successful dreamer had listened to them, no one would reach those coveted accomplishments. Ignoring the fears, doubts, and all the possible reasons *not* to pursue your calling is a prerequisite for success.

Let's consider children once more. This is a story that friend and motivational speaker Geo Derice shared with my performance-training center, House of Strength, a few years ago.

He called it *Fail Like a Child*.

I'm paraphrasing to shorten it, but think about a child learning to walk or ride a bike. They never succeed on their first attempt. Or their second. They probably fall on their third and fourth attempts also. And even after their first successful attempt, there will be more falls and slip ups, but those missteps never dampen the flame or dissuade their commitment to the pursuit.

At no point do they say to themselves "Man, I suck," or, "Wow, I'm a failure." Nor do they say, "I can't walk," or, "I'm not cut out to walk."

Like a child who hasn't yet developed the ability

to walk, each of us is only deliberate practice shy of whatever threshold we desire for ourselves.

Something else children don't do while acquiring these skills is compare themselves to others. A human learning to walk doesn't worry about what an outsider thinks about their level of development; they just continue to fight for every ounce of improvement, day after day, attempt after attempt.

Why do adults do the opposite around pursuits like health and fitness, building a business, writing a book, curating a blog, or public speaking?

This leads to something experts call "comparison inferiority complex." If children let the failure of their first few attempts or the opinion of outsiders affect their commitment to the mission, they would never learn to walk, write, or ride a bike.

As you go through this section, keep in mind the habits of the highly successful (think Elon Musk or Richard Branson) and our innate child-like ability to pursue progress.

How would these high performers respond? How would they spend their time? How would the highest version of you handle certain situations?

F*ck Your Feelings. F*ck your fears and doubts. F*ck the (mostly fearful or envious) opinions of outsiders and move as boldly, passionately, and confidently as a child.

Act *as if* you were that person you wish to be. Be and become the highest level of yourself.

The Magic Happens Outside of Our Comfort Zone

Your life is perfectly designed for the results you're currently getting. – Paul Reddick (mentor)

Let's talk about acting out of character. Comfort is the enemy of motivation. The less comfortable we are with our situation in life, the more impetus for action we have.

If you're not making the strides you want to make in life, it's because you're too damn comfortable in your current situation. If you were more uncomfortable, you'd have a higher-level motivation to drive change.

As the saying goes, "The hungriest man eats," with level of hunger being the discomfort in that maxim. A desperate survivor will do anything to preserve life – his/her own and that of their family. This includes any number of actions that, under normal (comfortable) circumstances, would be considered taboo or opted against.

From my own experience, when I started House of Strength in 2012, I literally would not have had the money to pay rent or buy groceries if I had not cold-called potential clients, sold memberships, and grown my roster of members.

I'd be lying if I said I have the same sense of

urgency today. I'm more comfortable with my financial setting and less driven to make those efforts – still driven, but not as much. There is something to be said for the inverse relationship between comfort and motivation.

I think this is a crucial point to keep in mind as we move through our journeys. Periods of stagnation are generally, at least in some part, created by comfort. If we find ourselves frustrated, stuck, or any other form of faltering, we should audit our comfort levels and investigate the real reason we're not as motivated (aligned) as we'd like to be.

This is why I like to seek adversity regularly. It builds resiliency into our character and spurs growth. Get used to disrupting your comfort level. Get comfortable being uncomfortable.

Acting out of Character

Acting out of character is a habit that can help us build resiliency, expand our comfort zone, and develop the trait of being a growth-seeking individual.

The more adversity we experience, the more resilient we can become. This is a crucial component of Navy SEAL training to develop the world's toughest warriors. Training is supposed to be harder than the real deal, otherwise, how would they be prepared to handle themselves with poise and clear thought when it mattered most?

We can apply this inoculation training to our own pursuits through steadily acting out of character, seeking adversity (within reason for our own goals), and expanding our comfort zones, in order to continually elevate our level of performance.

This could be an introvert joining a networking group or Toastmasters, ultimately working toward speaking on stage. The example matters less than the lesson itself. These pursuits of growth serve the important purpose of positively changing traits through practicing discomfort. Our default setting is to do what is easy, or comfortable. Unless we practice and train ourselves to do otherwise. The mind and body seek homeostasis. This is a dangerous thing for the person interested in perpetual growth.

Challenging our comfort zone, however, is a self-perpetuating habit. Doing things that scare us or are uncomfortable prove to ourselves that we can do it, empower us, build confidence, and create momentum. The more we do it, the better we get at doing it.

Getting Outside of Your Comfort Zone

Simply put, growth happens outside of our comfort zones.

By definition, if we have a skill or knowledge of something, it lies within our comfort zone. Maybe on the border if we are yet to master it, but the point remains: the skills, traits, and qualities we wish to possess but do not currently have lie outside of our current limits. It is only by pushing ourselves beyond those boundaries that we can learn, experience, and achieve new things.

For this reason, I am a huge fan of the phrase, "Get comfortable being uncomfortable."

I like to build as many comfort-disrupting habits into my life as possible. Things like cold showers, intermittent fasting, and challenging workouts are easy to implement on a daily basis.

Operational capacity is always lower than absolute capacity.

Systems (and people) cannot operate at full throttle for very long. Think of this as redlining a car's engine, or pushing a factory's assembly line speed of your one-rep max in the weight room.

If you could lift the weight two to three times, it wouldn't be your ceiling, absolute max, or your one-rep max. That is the definition of absolute output ceiling.

The rate at which we operate is somewhere below that ceiling. It's usually a comfortable-to-borderline stressful pace that is about 70-80% (not scientific number) of that absolute max. I call this "operational output."

However, if we continuously elevate our absolute capacity, our operational capacity (say, 70% for an example) will also increase.

Let's use the weight room for example. If my bench press max for one rep is 200 pounds, then 70% is 140 pounds. If I increase my bench press max to 300 pounds, then the same 70% output is now 210 pounds.

I'm still working at the same exertion level, but I'm using a greater load, doing more work (W = Force x Distance), and I'm able to build more muscle and expend more energy (burn more calories) than if I were still using 140 pounds.

This is an oversimplification of the strength and conditioning game, so for all my strength coach friends, please realize this is just an example. Here's another example in a different domain: speed reading.

If my maximum words per minute is 400 words, then my comfortable pace is closer to 250-300 words per minute. It's also likely that I lose comprehension at 400 words/minute. But if I increase my absolute ceiling to 600 words per minute, my comfortable (operational) pace will also increase.

Let's say it's now 350-400 words per minute. I can now consume about 25-30% more information in the same amount of time – *and* retain the information.

This concept of expanding our limits, broadening our comfort zones, and increasing our "operational output" is paramount to becoming the person we want to be – in all aspects of life.

We know life gets more and more complicated as we age. But our goal as humans is to become better equipped to handle bigger (and a wider variety of) stressors. This can only be accomplished if we constantly expand our comfort zone.

As my friends at Atomic Athlete in Austin, Texas say: "Be an asset, not a liability."

Need a more science-based reason to expand your horizons? Novelty up-regulates neuroplasticity.

Neuroplasticity is the brain's ability to create new neural pathways. These new pathways can be involved in healing, learning, and making new connections; most importantly, they can continue late into life – one reason they're shown to have the potential to significantly reduce risk factors for cognitive decline.

"You should think of a real life as one in which you are continuously learning," said Dr. Michael Merzenich. Dr. Merzenich is regarded as the father of neuroplasticity, and he's the founder of BrainHQ, a mental gym of sorts that utilizes the science of neuroplasticity to help members think faster, focus better, and remember more. [132]

When I interviewed Dr. Merzenich, he used the example of learning to read to demonstrate neuroplasticity and the brain's ability to change and create itself.

We're not born with the ability to read. We don't have a "reading brain" before we learn to read. We must create it from scratch. The reading brain must be developed step by step, from learning the alphabet, sounds, words, sentences, and into more complex abilities like context and comprehension. We create the ability to use these visual and mental skills to be ability to glance at a paragraph and quickly understand the communication. If we never learn to read, we don't have this reading brain.

As Dr. Merzenich explains, every time you learn a new skill at a refined level, you create a controller in the machine (your brain) for that ability. Novelty and new experiences – any stimulating input – feed this neuroplasticity cycle and facilitate positive brain evolution. We do this continually in life, and the more we do it, the more powerful our brain and the more complex a person we create.

Peak Brain Institute's Dr. Andrew Hill has confirmed this as well, stating that "novelty upregulates neuroplasticity." The more new experiences we exposure ourselves to, the more our brains develop and grow. Be a lifetime learner – for brain health and human evolution. This brain development habit is a microcosm for personal development. The more exposure we have to experiences outside of our comfort zone, the more we can expand and evolve as humans.

The exciting part of neuroplasticity – and this has been verified by both of these experts and a growing body of research – is that we are not relegated to suffer from the previously thought of as inevitable age-related cognitive decline. If we continue to challenge our brain throughout life and ask it to grow, it will resist age-related decline. (There is also a dietary component to this, but that would be its own book entirely. Check out *Deep Nutrition* by Dr. Cate Shanahan to see why sugar, vegetable oils, aka "liquid age", and the associated oxidative stress are public enemies number one and two.) [240]

Tips for adding new skills and expanding your comfort zone

Without talking to you and identifying blind spots or weaknesses, I'm not able to highlight the specific areas that need your focus. You'll have to identify them yourself using the reflection questions on the ensuing page.

I can (and will) offer suggestions, but if you don't find them stimulating, you're not likely to be consistent with them. Only you know what interests you. Some of the most popular ideas seem to be learning a musical instrument, speaking a new language, picking up a new trade craft, or learning a career-enhancing skill.

You don't have to limit yourself to one domain. Using myself as an example, I want to learn and experience it all. I'm equally comfortable talking biochemistry with leading neuroscientists, as I am smashing weights with National Champion Strongman competitors. I'm just as comfortable hunting big game in remote wilderness as I am walking a runway at New York City's fashion week.

None of those things would have been possible without my efforts to expand my horizons, so I encourage you to follow your own passions and realize that there are no limits to what you can achieve.

"Grow or die." This is another lesson I took home from my SEALFIT experience in 2012. In nature, the moment we stop growing is the moment we start dying. Look at the flower that reaches peak bloom, then withers and dies. Or look at the Alpha male lion who, once defeated and relieved of leadership duties, is sent off to slowly decline into death.

This is the natural order of our world. To stop growing is to start dying.

Never stop growing. Or as the SEALs say, "always sharpen the knife."

Reflection Questions

1. Your life is perfectly designed for the results you're currently getting. What results are you happy with? What results would you like to get that you are not currently getting?
2. What changes can you make in the way you spend your time in order to get more of the results you desire?
3. How are you holding yourself back?
4. What would you do if you knew you could not fail?
5. Ask yourself: what is the fear? (credit: Paul Reddick, back in 2012-2014)
6. What ruts are you in because of your preference for comfort?
7. Where can you seek adversity, disrupt your homeostasis, and expand your comfort zone?

Activities

Comfort-Zone Challenges

Given what we've discussed about comfort being the enemy of motivation, I like to seek adversity as often as possible. I like to constantly challenge my comfort zone in an effort to continuously expand and evolve as a human.

Our Navy SEAL friend "Ryan" likes to seek adversity daily, sometime leaving his phone, keys, and wallet at home, then running as far as he can in one direction and forcing himself to find his way home. Other days, he stand up paddle boards six miles to work instead of driving.

Here are some ideas for daily, weekly, quarterly, or annual challenges to spur growth and expand your comfort zone. Generally speaking, daily challenges will be smaller and shorter than quarterly or yearly challenges.

<u>Daily</u>

- Cold shower
- Meet someone new/talk to a stranger
- Perform a random act of kindness
- Walk backwards on a public sidewalk (not a busy one)

- Hold a power pose in public

Daily challenges should be designed to help inoculate you against stress and fear and get you more comfortable being outside your now ever-expanding comfort zone. That's the thing about comfort zones: the more your push them, the larger they get.

Weekly

- Join Toastmasters and work on public speaking
- Do a once-weekly challenge workout to test mental and physical toughness
- Learn a musical instrument (e.g., weekly guitar lessons)
- Learn another language
- Enroll in a jiujitsu, boxing, or other class

Weekly challenges should be deigned to help us learn new skills – guitar lessons, jiujitsu classes, weaponry, or carpentry. Always look for ways to increase your value as a human. Continue to grow, expand, and be more of an asset.

Quarterly

- Spartan race (or other obstacle course race)
- Three-day technology detox (no computer, cell phone, or TV)
- Five-day water fast (huge health benefits and resets relationship with food)

<u>Annual</u>

- Physical competition – powerlifting meet, bodybuilding show, strongman contest, SEALFIT – something you have to train for over an extended period of preparation
- Week-long backpacking or hunting expedition
- Run a marathon or ultra-marathon

I really like the idea of quarterly or annual "gut check" events. Run a marathon or compete in a Strongman event once a year to prevent stagnation – or worse, regression. Do SEALFIT, a Spartan race, enter a bodybuilding show or powerlifting meet – there is tremendous benefit from simply signing your name on the registration list. You're now committed, and your name/ego/pride are on the line. Most people respond favorably in such situations. When in doubt, do the thing that scares you the most. Each of these challenges has its place, and I encourage you to implement both short-term and long-term challenges to your comfort zone.

Breath Work

I separated breathing from yoga and meditation because I find breathe work to be a more powerful and immediate state-changing tool than yoga and meditation.

Breathing is also the easiest, most convenient, do-anywhere tool we have. You can do it right now without stopping your reading (or listening). You can do it within a workout, while working on the computer, driving, or on a plane/subway without disrupting your neighbor. We can't say the same thing for yoga or burpees.

Each has the ability to enhance our long-term traits, but I think (and have seen) breathing change more people's immediate states than either yoga or meditation.

Why breathing works

Besides supplying the oxygen vital for life, our breath controls the autonomic nervous system via the vagus nerve, which is made up of the parasympathetic (PNS) and sympathetic nervous systems (SNS).

HRV, as we've discussed, is a major indicator of how much time we're spending in parasympathetic or sympathetic states. By definition, we can only be in one at a time. So PNS activities, like meditation, shut down the SNS and vice versa. More time in the PNS is correlated to higher vagal tone (or HRV), whereas more time in SNS activation is correlated to decreased vagal tone (HRV).

The vagus nerve, as we've discussed, is made up of afferent and efferent pathways, making it a two-way street. It both controls and is influenced by our breathing.

Think about fight or flight scenarios where breathing is shallow and rapid. This type of breathing either signals or is stimulated by fight or flight situations. We want to be aware of our breath so that we're not inadvertently sending our body into sympathetic states simply because we have dysfunctional breathing patterns. As you'll hear in a later section on posture, most people are guilty of this.

Conversely, think about relaxing experiences (like a zero-gravity message chair before entering a float tank) and how breathing is deeper and slower, stimulating that relaxed, parasympathetic state.

Again, more time in the parasympathetic state increases vagal tone and emotional resiliency. This is what we're after with our long-term, trait-developing habits.

In the long-term, big picture view, we need to work to spend more time in these states, and since we're constantly breathing, this is a very effective place to start. It has compounding, long-term benefits, but breathe work also has profound impacts on our immediate, short-term, transient states.

How to implement

My favorite ways to implement breathe work are box breathing and what has become known as the "Wim Hof Method."

I wish I had found breathe work earlier in my life. It was 2012 when I first realized its power and importance. Since then, it has completely changed my ability to meditate, my yoga practice, how I deal with stress, how I recover from workouts, and almost every aspect of my life.

As we prepared for SEALFIT that September, I was exposed to many of Mark Divine's lessons, and box breathing was one of them.

I like box breathing best for the simple fact that it can be done anywhere. The Wim Hof Method, as you'll see, usually requires the person to close their eyes and lay down. This makes it difficult to use in traffic, while walking, or in many real-world situations. For this reason, I use Wim Hof as a once-daily, long-term practice, while incorporating box breathing throughout my days for instant boosts or fixes.

I learned the box breathing method from former Navy SEAL commander and founder of SEALFit, Mark Divine. Mark teaches a four-phase breathe that includes an inhale, a hold, an exhale, and a final empty hold. The key to box (or square, as in all sides equal) breathing is that all these phases of the breath are equal in length. [241]

You can use whatever tempo you like 2-2-2-2, 4-4-4-4, 5-5-5-5 – just make sure all phases of your breath are equal in length. Deeper and slower than normal breathing should be your goal.

Typically, I prefer the 5-5-5-5 because this makes one cycle last twenty seconds, which means three of those cycles equal one minute. I can then perform three to five such cycles without looking at a clock and get three to five minutes of mindful breath work which can also be paired with meditation.

I also find that I prefer to "count" my fives with heartbeats rather than seconds. I place one hand over my heart and inhale, hold, exhale, and hold again – all to a count of five heartbeats. Again, this frees my mind up from having to count or worry about that man-made construct of time.

1. Inhale x 5 heartbeats
2. Hold full breath x 5 heartbeats
3. Exhale x 5 heartbeats
4. Hold empty x 5 heartbeats

5. Repeat for 3-5 cycles.

When I interviewed Dr. Stephen Porges, he recommended making the exhalation longer than the inhalation to promote increased vagal tone and recovery. Since that conversation, I've adjusted my "box" breathing to have an extra one or two counts on the exhalation, so a 5-5-5-5 now looks like 5-5-6-5. [82]

When should we box breathe? Anytime.

I will use it any time I am feeling stressed, when I am driving, or when I am working. I try to remind myself several times a day to be aware of my posture and my breathing. Use it as a start to your meditation practice – one of the fastest ways to get into a meditative state is to focus on your breath – or use it in the morning as part of a grounding routine to start your day or at night as part of your sleep routine.

Box breathing, or any focused, intentional breath work, is perfect for the "in-between" moments. Driving, walking to another room, waiting on a call or waiting in line – anytime you're not actively engaged could be considered "in-between" times and a great opportunity to focus on your breath and center yourself. Even between sets in the weight room, although sometimes the tempo has to come down to 2-2-2-2, and I work to move it toward 3-3-3-3 by the end of the rest period.

The second breathing method I want to share is the Wim Hof Method. In short, the practice is three to four rounds of thirty breaths where you focus on the inhale, trying to get as much oxygen into your body as possible.

This state of hyper oxygenation can lead to a

tingling sensation and sometimes lightheadedness, which is why it's best to practice this lying down, not driving in rush hour traffic.

There are entire books on this, so please seek them out if you want to learn more about this method. Google Wim Hof, visit his website, or attend a seminar where he or his staff teach the method.

Wim's method is built on the ancient Tibetan Buddhist practice of Tummo, or "inner fire." Originally cultivated for tantric breathing cycles and yogic heat, Tummo is now used as the breathing element of Wim Hof's famous cold exposure. [242, 243]

Cold Exposure

The warmth may be comfortable, but the cold is your friend. – Wim Hof

The easiest way to implement cold exposure is a daily cold shower.

I love the concept of beginning each day with the discomfort of a cold shower for the psychological benefits. As we've just pointed out, it's stress inoculation. We're voluntarily thrusting our comfortable selves into the uncomfortable cold water and consciously rewiring our primitive urge to flee.

If we start our day with a cold shower, we notch that first victory, creating mental resolve that can never be taken away from us. Knowing that you voluntarily did something uncomfortable, something that most would run from, is empowering in its own right.

Couple that with the norepinephrine hit, the increased vagal tone, and the stress inoculation we just discussed, and you've got yourself a powerful tool.

Most of us shower daily already, so switching the water temperature from hot to cold isn't a great imposition on your schedule – you may just find that you take a shorter shower. You'll also build confidence the more you do it, you'll expand your comfort zone and learn to become comfortable being uncomfortable.

If a cold shower is too much right away, you're missing the point. They're cold showers, they're *supposed* to be uncomfortable. That's the whole point and why they provide so much benefit.

However, you can try contrast showers, where you alternate hot and cold water. Start with twenty seconds of hot, then ten seconds of cold water for five minutes, eventually working your way to 15/15, then twenty seconds cold and ten seconds hot.

Personally, I don't like this method. In the number of days it takes you to work up to full cold showers, you would have already acclimated to straight up cold showers and enjoyed the psychological and metabolic benefits listed below.

You can also dunk your face in ice-cold water. And for those who want the full immersion at home, you can do ice baths or look into building your own cold tub. Consider yourself warned: full body submersion in ice water is a significantly different experience.

Any farm store sells large steel troughs that can be filled with water and ice. There are "how-tos" all over social media where people bought small freezers and built boxes and stairs around them and figured out how to keep them at 40 degrees. Do a search for "convert your freezer to a cold tub."

There is also cryotherapy, which is becoming so popular that almost every city has a location for you to try this. And depending on where you live, simply walking outside in the winter is cold exposure enough. (Again, don't get frostbite or die of hypothermia, please).

The goal with all of these exposures should be to breathe deeply and slowly. You want to get past the shivering, and into the NST (non-shivering thermogenesis) state for the BAT (brown adipose tissue, explained below) and parasympathetic adaptations.

When we were in Finland, alternating jumps in the frozen lake with Finnish saunas for heat exposure, Dr. Rhonda Patrick explained to me that usually five minutes at below 40 degrees is sufficient for cold exposure.

What I love most about cold thermogenesis is that you don't have to go to Finland to jump in an icy lake. Like breathing or meditation, it's a tool that is readily accessible for almost everyone. Like meditation, breathing and cold exposure are potent daily practices of controlling our nervous system and limbic defense responses.

Cold: benefits beyond your mind

Cold exposure has numerous benefits outside of our conversation on vagal tone and HRV. Entire books have been written on this, so for the sake of brevity, here are the highlights. [244, 245]

Cold exposure has been shown to help convert white fat to metabolically active brown fat. BAT (brown adipose tissue) actually uses fat for fuel rather than storing it. [246, 247]

The first study found that ten days of cold exposure was all it took to increase BAT activity in adult humans and increase NST (non-shivering thermogenesis), a process that increases core temperature without shivering. This is done through sympathetic nervous system activation of the BAT. [248]

Another study, albeit in rats, found that cold exposure revamped thyroid function and increased BAT oxygen consumption by 450%. [249]

Two of the leading experts in both heat and cold exposure are Dr. Rhonda Patrick and Dr. Jack Kruse. Both are intelligent truth-seekers who do a wonderful job of assimilating research and distilling it down to useful chunks of information that we can implement.

According to their writings, cold exposure has been shown to increase norepinephrine (up to fivefold increases in the brain), boosts mood, increases focus and attention, increases cold shock proteins that prevents atrophy and repairs damaged synapses, lowers inflammation mediators, may boost immune cells (including anti-cancer cells), increases mitochondria count and efficiency, and promotes fat burning. [10,11] [250, 251]

If you seek more information on breathing, heat, or cold exposure, I strongly encourage you to follow up with Wim Hof, Dr. Rhonda Patrick, and or Dr. Jack Kruse – links will be provided in the resources section of this book.

CHAPTER 8:
Master Your Actions. Control Your Future.

Faking It

Fake it till you make it. – Tony Hart *(my high school basketball coach)*

I hated that damn quote. At sixteen, I had no idea what it meant. It just sounded stupid. Admittedly, I ignored it. He said it so much that I became desensitized to it. But here I am, nearly twenty years later, hearing it in my head as if I were at practice last night.

Anyway, Coach Hart was always talking in quotes that we (read: *I*) didn't get. For example, one weekend after a Friday night loss, he gave us the homework assignment to spend the weekend meditating on this gem: "Discretion is the better part of valor."

I remember actually trying to understand it, but come Monday's practice, none of us had an explanation that satisfied coach. Looking back, I realize what a tall ask it was for a teenage male to understand such a powerful truth.

I don't think I fully understood it until I was twenty-seven or twenty-eight. And now, a few years on the north side of thirty, I'm only now learning to embody it.

Back to faking it until we make it, it turns out that Coach Hart was teaching us the power of the mind at a very young age. What he really meant was "your imagination is a preview of your future."

He was teaching us that we had the power, in our own minds, to create whatever reality we desired for ourselves. Since 95% of our decisions, and therefore our actions (or inaction) are based on how we feel, it's vital that we devise a repeatable system to ignore and overcome our feelings before they can control us, hold us back, and sabotage our best laid plans.

According to coach, this meant faking it until you make it.

I like to call it "Act As If."

They say you should dress the part of the job you want. I say take it a step further, and act as if you already have that job. I've always told my members and client to act as if you were the person you want to be.

For example, envision the person you want to be in three, six, or twelve months. Or if helps, envision the person you respect, admire, and look up to. Maybe it's Tom Brady, Richard Branson, Elon Musk, or Peter Diamandis.

> How would that person spend their time?
> How would that person act in this situation?
> What does their day look like?

Once we've thought about and visualized how they would spend their time or move through life, the trick is to begin to implement that and make it how we spend your time *right now*.

This exercise teaches us to reverse-engineer success and create that which we want to achieve. It also uses the science of neuroplasticity to help us rewire our brains, learn new ways, create new habits, and become the person we want to be.

Sometimes it's hard for us to see ourselves objectively. In these instances, it can be useful to observe successful people and model their behaviors. Take Tom Brady or Peter Diamandis for example. What if Tom Brady came back to the huddle after a bad play and said, "Well guys, we suck, we're screwed." Odds are, a quarterback with those poor leadership skills wouldn't have the career Tom Brady has compiled.

What if Peter Diamandis said, "The game is rigged, my fate in life is out of my hands, I blame the government, and I'm checking out." Again, odds are, a person with that mindset and those actions isn't going to enjoy a fulfilled, satisfying life.

Act As If is a great tool to audit our behaviors and maintain alignment. The fastest way to become the version of ourselves that we most want to be is to start acting as that person in every possible moment.

Would they hide from it?
Would they research it?
Would they employ a trusted team of support or would they handle it on their own?

Do you see how your past handlings of these hypothetical scenarios are not congruent with the actions of the person you want to be?

If so, this is a powerful revelation. It shows us exactly how and where we went wrong previously – but more importantly, it shows us the version of ourselves that we want to be.

The narrative we hold in our heads about who we are determines what we will accomplish in this life.

It's not who we are that holds us back; it's who we think are not. If you know you're capable of more, here's a quick technique to flip the script in five minutes.

1. List three words that define you *now*.
2. List three words that define the person you want to be.
3. Set a reminder on your phone with the three words that describe the person you want to be. Have it go off at least once per day to serve as a reminder to Act As If.
4. Act as if. Live those three words. Become that person

Time Management

We are what we repeatedly do. Excellence then is not an act, but a habit. – Aristotle

Remarkable people do unremarkable things with remarkable consistency. – Logan Gelbrich

This chapter is all about our values, our actions and how we spend our time. People don't get results. Processes do. When our time here in this world is up, we'll be defined by how we spent that time.

Here are some tools to optimize how we spend our time and prioritize our actions.

Time Management and Productivity

When I use the word "productivity" in this section, I don't necessarily mean it the way we see it used so often today. It's not about how much stuff you can cram into your to-do list or your week. Don't make the all-too-common mistake of confusing being busy for being productive. I'm using the term in the sense of making the most of our time here in this life, on this planet.

Watching Netflix may be a great way to decompress and relax, but it's not going to help us create lasting impact on the world (unless you wrote, acted in, or in some other way created that show that adds value to others' lives).

If you haven't picked up on it by now, our mortality is never far from my conscious mind. I wouldn't say I'm afraid of death, but I will say that I am incredibly aware that my time here is finite. And ceasing to exist scares the shit out of me. Specifically, the possibility of ceasing to exist before I experience and accomplish the things I want to accomplish fuels me to waste as little time as possible and to make as much impact as possible while I can.

When I talk time management or productivity, and when I weigh the pros and cons of certain activities, this is the lens through which I'm viewing things: are we honoring our existence? Can we do more? Can we be more? Is this the best use of my time, skillset, or expertise?

From our 50,000-foot view, our habits are the climate (versus weather) of how we spend our time. Success is not an accident. We either have habits that create success or we don't.

The good news is that with the increased awareness and consciousness that we've worked to develop so far, we can consciously choose to engage in habits that foster success. We can be in control.

Aside from our health, time is our most precious commodity. We need to realize how finite and precious our time here on earth really is. In everything we choose to do, we're exchanging our time for that activity. I believe life is too precious to waste time doing things we aren't passionate about or with negative people.

I actually agree with the memes on the Internet, when faced with decisions, we should say "f*ck yes" or it should be a no.

Before we look at the power of saying no, allow me to interject the requisite disclaimer or caveat now. Every endeavor involves struggles, growth requirements, and attacking our weaknesses; do not confuse these with "things you hate" and quit or give up.

When I owned my own gym, I had to plunge toilets, take out garbage, and do many other things that weren't my favorite. There will always be aspects of a career or pursuits that are less desirable. These are unavoidable, but they're required if we're to spend our time (and our lives) doing what we love. They're part and parcel with the journey. Understand this difference.

High-Performance Principles of Time Management

Nothing is going to execute itself, and hope is not a strategy.

There is no secret sauce or magic red pill to "hack" time management. Like everything in this book, it's about awareness of where we direct our attention and how we spend our time. It's about having a plan and sticking to it. Those who succeed in managing their time effectively do not do so by accident.

As its core, time management is understanding what to do, when to do it, and who should do it.

Habits are defined as settled or regular tendencies or practices. They're choices, and as Aristotle pointed out, they're the way we repeatedly choose to spend our time.

Our lives are busier than ever. We're constantly bombarded with twenty-four-hour news, sports, and weather. Our smart phones have us connected to social media, email, texts, and phone calls all day, every day.

The average American spends 7.7 hours per day sitting, 10 hours per day looking at screens and as of 2007 (pre-social media explosion) consumes 174 newspapers worth of information daily.

No wonder it is so relaxing to simply sit, observe nature, breathe, and ignore your phone for a few minutes.

Interestingly, University of Michigan researchers have shown that twenty minutes in nature reduce stress and improve brain function. Even looking at pictures of nature works if you're not able get outside.

I'm not advocating a complete regression into

the Stone Age. Although many of us could benefit from a short-term digital detox, dropping off the grid is not realistic for most. So how can we find balance and effectively manage our time and schedules without sacrificing our health, sanity, and productivity?

That's exactly what I'm about to share with you here.

I've spent the last few years working with and interviewing neuroscientists, behavioral researchers, and elite performers to deconstruct the habits that help them achieve greatness.

The results are what I call the neuroscience of high performance and self-mastery.

The short answer is this: by bringing more awareness to where we focus our attention and how we spend our time, we can regain some sanity, reduce mental clutter, lower stress, and be effective in every area of our lives.

The slightly longer answer, and your blueprint for high performance management, is below.

Nine Essentials for High-Performance Time Management

1. Don't check your phone first

How long are you awake before you check your phone?

46% of Americans (and 66% of millennials) confess to checking their phone before getting out of bed each morning, and one-third admits to waking up and checking their phones during the night. [252]

I understand the desire to check in with the outside world. But doing this before our feet ever touch the ground is a dangerous habit – and not one engaged in by high performers.

High performers and successful people usually go to bed with a plan for the next day. When they wake up, it's all about executing that plan, not wondering what other people are posting on social media or what fires await in their email inbox.

Do we really think Richard Branson or Oprah wake up and check the number of likes on their status updates or allow their email inbox to dictate their morning? They don't.

High performers handle their own shit first, then worry about others. Set a "check in" time that will be the first time each day that your turn your phone on. Up until that point, the day is yours. Spend your time on yourself, with your family, or being productive on a big goal or a side project. This simple rearrangement of priorities and schedule will do wonders to boost productivity and happiness.

2. No unscheduled phone calls

The moment we decide to stop what we're going to answer an incoming call is the moment we decide (consciously or subconsciously) that the incoming phone call is more important than our current engagement.

We're letting that person or that note highjack our mind, our energy, and our attention. This may sound harsh, but let's take a closer look at an all-too-real example.

Consider this email that many of us might find in our inbox one morning.

Hi Ryan,

We're out of your awesome product. Please send us four cases as fast as you can, we have customers screaming for more.

Thanks,

Valued Business Partner

While this appears to be a politely written email, any savvy time management ninja can see what this is really saying.

"My problem is more important than whatever you're working on. Please stop what you're doing and fix my problem."

This customer's failure to properly identify inventory levels and predict lead times for purchases is their mistake. It's unfortunate, and I will rush that order to them — that's what you do for a friend, but only when it's time to check emails and respond to them.

If I were to check this email before getting out of bed, this person would successfully have inserted their agenda into the top of my to-do list.

These interruptions and distractions are what *you* open yourself up to by checking emails or answering unscheduled phone calls, and it is perfect example of allowing other people's stuff to derail your maximum output sessions.

I love our customers — they are what keep businesses running. But there is a proper way to manage your time for maximum productivity, and being at the 24/7 beckon call of hundreds, thousands of customers is not the answer.

I rarely answer my phone for unscheduled calls, and I never answer calls from unknown numbers. I have a code that my family knows, so they can get me to answer if need be during the work day. Otherwise, it's text and wait for a reply (not immediate).

If we're on the highway in the midst of an eight-hour road trip, this could be a welcome interruption. However, if we're drafting a sale pitch, creating content, or doing any other high-value activity, allowing this interruption is a terrible time-management decision. More on this in the quadrants of importance below.

When we're doing our high-value work, we should not answer incoming phone calls, check email or allow any other distractions.

In fact, for maximum time management, most high performers schedule all phone calls and vehemently refuse to answer any unscheduled incoming calls. This keeps them in control of their most precious assets: their time and their attention.

Pro tip: when you're being productive on your computer and don't need your phone, keep it on airplane mode.

3. Only check emails at predetermined times

Email can be one of the biggest time-sucks in existence. Here's your strategy to spend less in your inbox.

Set two or three times each day to check and respond to emails. 10am and 4pm are solid choices. The mid-morning check gives you time to be productive before the emails fly, yet still allows you time to respond the same day if need be, while 4pm does the same for the afternoon.

If you feel compelled to reply instantly, set an

auto-reply message that informs people of the times that you check emails so you can set the expectation for when they can expect a reply.

When you do check your email, use the three Ds from Mel Robbins, author of *The 5 Second Rule*: [253]

1. Deal with it
2. Delegate it
3. Delete it.

Finally, don't keep your email tab or window open. Close it. It will still be there when it's time to check email again in six to twelve hours.

4. Pomodoros: literal time management

Pomodoro is the Italian word for tomato, and the Pomodoro timer is a method of breaking up your work into smaller, more manageable chunks of time.

Traditionally, they're twenty-five-minute windows of work followed by five-minute breaks, repeated four times for a total of two hours, then followed by a longer, thirty-minute break.

1. Twenty-five minutes of focused work
2. Five-minute break
3. Twenty-five minutes of focused work
4. Five-minute break
5. Twenty-five minutes of focused work
6. Five-minute break.
7. Twenty-five minutes of focused work
8. Five-minute break
9. Thirty-minute break

Minding the timers is crucial. The power of Pomodoros is lost if you are not militant about stopping and starting when the timer dings – no finishing an email or paragraph before stopping for your break.

They are a surprisingly powerful technique with several reasons they work to help increase productivity and quality of work.

There is something about seeing the clock count down that increases focus and motivation. I can't promise you'll feel this, but I know it motivates me to see how much I can get accomplished before the twenty-five minutes elapse.

Pomodoros allow you to keep your brain fresh.

Mental processing, like muscular contractions, produces a metabolic waste by-product. If we let these waste products build up, we get headaches, or worse, reduced brain function, so we need these periods of recovery.

Think of it as avoiding burnout, neurotransmitter depletion, or brain fatigue – like doing 3x10 in the gym instead of 30 reps straight through. We know form would break down in that instance; our brain is the same way.

Using Pomodoros allows us to stay fresh, avoid fatigue, and to produce higher-quality work with a greater output.

You can set a timer on your phone or do a quick search on the internet for "free pomodoro timer" to find a desktop version.

The traditional 25-5 setup is not the only way to use this time management principle. I also like to use fifty-minute blocks of focused work with a ten-minute break to finish every hour. I go outside and walk 500-1000 steps at fifty minutes after every hour. This utilizes the Pomodoro technique to keep my mind fresh, but also reminds me to move frequently and gets me outside in the sunshine and in nature. It's a true "stacking" of effective habits (see Rule #8).

5. Limit distractions

This one is simple, yet often overlooked. If our phones are airplane mode, it's a non-issue, unless your computer has notifications turned on.

In that case, turn off your notifications. We don't need to know about every single email that lands in our inbox. If we're focused on a task and being productive, we don't need to know that someone commented on our last social media post.

Turn off notifications, work when we're charged with working, and compartmentalize all other tasks, tools, and apps.

Remember this: flow follows focus. Those much sought-after flow states that allow for peak performance require focus before flow can be experienced.

Stanford research agrees. A 2009 study on multitasking demonstrated that multitaskers have less control of their memory and "are slowed down by irrelevant information." [254] They continued, saying people who engage in multitasking are *"suckers for irrelevancy."* (Their words, not mine. But if the shoe fits...)

If you work in an office, notify your co-workers of your preferences. People with your (and the mission's) best interests in mind will respect these boundaries.

6. Have a plan (and stick to it)

Effective time management, like healthy eating, does not happen by accident. It requires planning and intentional action. But even the best plans are worthless if they're not followed.

Recall Damasio's finding that 95% of our decisions are made based on feelings, not logic. This explains (at least in part) why 70% of people who set goals fail to reach them. [255]

As we've said all along, do it anyway. Act as if you were the highest version of yourself. How would that version of you handle this moment?

Achieving goals or being effective with time management require us to chart a course of action, like a map that takes us from where we are currently to where we want to go. This map is only the first part of the equation. We must also show up, walk the path, and stick to the plan, no matter how we feel on any given day.

The ability to do this is something that sets high performers apart from the rest. High performers know that processes, not people, get results.

Caveat: some days we need to deviate from the plan, which can be successfully accomplished if you understand the Move the Chains concept.

7. Move the Chains

This is the same concept we outlined in the Preface. Move The Chains. Make progress.

I try to do at least one thing every day to make

progress toward my goal(s), no matter how small that act may seem.

The crucial lesson behind the Move The Chains concept is not to worry about flashy, fancy plays. It's not about the Hail Mary passes or the highlight-reel plays. It's about doing the boring, mundane, yet *effective* things relentlessly to constantly move our mission forward.

Identify something each day that equates to the proverbial 3.4 yards for your mission and make sure you're doing at least one thing every day that moves the chains.

More often than not, that single act gets the dopamine flowing and one act turns into two, then three and before we know it, we're crossing off half our to-do list – all because we got started.

8. Stack activities for multiple benefits with single efforts

While multitasking is ineffective and inefficient, certain activities can produce multiple benefits.

For example, a telephone call or meeting can be modified to be conducted while taking a walk outside in nature. Instead of accumulating more time sitting and staring at a screen, we're now up, moving, outside, and in nature – or at least a green space, if possible.

Another example could be signing up for improv classes or joining a Jiu-Jitsu class. In both cases, our brains are forced to learn something new, which up-regulates neuroplasticity (prevents cognitive decline), we get social interaction (release of feel good hormone oxytocin), some play/de-stress activity, movement, and time for ourselves.

Where in your life can you incorporate this concept of singular activities with multiple benefits for your health, productivity, and balance?

9. Understand (and employ) the quadrant of importance

We must not confuse being busy for being productive.

Stephen Covey's Quadrants of Importance, presented in his famous book *The 7 Habits of Highly Effective People*, is a visual concept that I found very helpful in determining order of importance for task. I think of it like medical triage. Gunshot wounds and heart attacks get treated before broken limbs, which get treated before headaches. [256]

	Urgent	Not Urgent
Important	**I** (MANAGE) • Crisis • Medical emergencies • Pressing problems • Deadline-driven projects • Last-minute preparations for scheduled activities **Quadrant of Necessity**	**II** (FOCUS) • Preparation/planning • Prevention • Values clarification • Exercise • Relationship-building • True recreation/relaxation **Quadrant of Quality & Personal Leadership**
Not Important	**III** (AVOID) • Interruptions, some calls • Some mail & reports • Some meetings • Many "pressing" matters • Many popular activities **Quadrant of Deception**	**IV** (AVOID) • Trivia, busywork • Junk mail • Some phone messages/email • Time wasters • Escape activities • Viewing mindless TV shows **Quadrant of Waste**

We need to know how to determine what is truly urgent. Until we master this skill, everything will seem urgent. Notice that most phone calls and emails fall in the "not important" and "avoid" quadrants.

Take some time to identify the task you need to complete and determine where each falls on this quadrant. Spend most of your time on the important items and delegate or drop the rest.

Notice immediately that the bottom two squares are waste and deception. Right there, 50% of tasks can be classified as waste or deception and should be avoided if possible – or delegated.

Understanding the difference between being

busy and being productive, and allocating our time accordingly, is paramount to becoming the person we want to be.

It's up to us to share and communicate these boundaries to the people in our lives and on our teams so that our productivity can skyrocket. In order to effectively implement this, we must identify high-value times and high-value activities and avoid time and energy drains.

Delegation is habit of every high performer I've worked with. CEOs, Navy SEALs, Olympic Athletes – they all understand that their success depends on them being able to focus on their specific duties, while allowing their teams to handle the rest.

Consider U2 and their self-proclaimed "Greatest Rock Show on Earth." Each concert venue along their tour required nine (yes, *nine*) days to set up and disassemble the show.

During this nine-day stretch, U2's Bono has one job, and one job only. That job was to show up and sing. He does not set up seating, sell tickets, worry about merchandise sales, or do sound checks in each arena. He simply shows up to the gig and performs his genius. Everything else has been delegated.

Make no mistake, ticket sales, merchandising, concessions, and sound quality are all important (even crucial) to the end user's experience and to the bottom line of the tour, but they're not Bono's specialty, and they're left to the capable hands of someone who specializes in those areas.

Imagine how much Bono's performance would decrease on stage if he had to worry about those other tasks.

Be like Bono. Identify your genius and delegate the rest.

Since I don't know the specifics about your situation, it's impossible for me to prescribe directives on how best to allocate your time. However, this is exactly what I do for the Olympic and professional athletes, CEOs and entrepreneurs I work with. If you're interested in working with me to cultivate higher performance as an individual or within your organization, visit ryanmunsey.com/consulting for more information.

Finally, some closing thoughts on social media, happiness, habits, and the use of negative space.

Time Management and Social Media

Amateurs Tweet. Pros Work. – Steven Pressfield

The world we live in is changing rapidly. Social media is a large part of this change, and it can be both a positive and a negative in our pursuit of the optimal life.

I've made countless friends and life "upgrades" through the connections that social media enables. Social media is becoming almost a necessity for brands and businesses, even offline brick-and-mortar businesses.

But when it comes to personal usage, we need to avoid the dysfunctional use presented in previous chapters. Our limbic system can easily become addicted to social media.

I'm not here to support or condemn its use. But

I will say any use should be intentional and done with awareness and purpose, not mindless consumption.

Here are some statistics to help make the decision.

An extra hour daily on social media increased children's unhappiness by 14%. [257]

More time on Facebook is linked to increased depressive symptoms. [258]

In fact, a separate study found a 1.7-2.7X greater risk of depression for those who spent the most time on Facebook. [259]

A March 2017 study from influencer marketing agency Mediakix calculated average time spent per day on YouTube (forty minutes), Facebook (thirty-five), Snapchat (twenty-five), Instagram (fifteen) and Twitter (one) and projected those figures out over a lifetime, arriving at a total of five years and four months. [260]

Just think of what we could accomplish in those five years of "wasted" time on social media. Combine the time we spend on social media and watching TV, and it's easy to see where we could find the time to make our own dreams a reality.

I know, because I've done it when I built House of Strength.

Forget the potential links to depression, forget the addictiveness; do you really want to spend five years of your life watching cat videos and other people's highlights? Take care of your own shit before you look at other people's shit. It's not that successful people have more time in their days/lives – it's that they spend/use their time more wisely than "unsuccessful people."

In five years, you could climb Mt. Everest thirty-two times, or fly to the moon and back thirty-two times. The point is, what are you trading, what are you giving up, what are you missing out on just to be on social media? Be aware of what you're doing and limit or reduce social media to ensure that it does not detract from your *real* life and real experiences.

Disconnect Completely and Be Unavailable at Times

Like an artist who understands the use of negative space, true experts of time management understand the power of no and time spent doing nothing.

I find it very important to disconnect from the outside and be present with my family. This is why I turn my phone to airplane mode every night from 9pm to 9am when I'm home. If I'm traveling, this window may shift a few hours one way or the other.

But generally speaking, when I'm home for the night, everything I need is right there. The outside world can wait. 9am may seem like I'm sleeping in. I'm not. I don't turn my phone on the second I wake up.

I usually spend the first two or three hours of each day with my phone on airplane mode, focusing on my morning routine and writing – whether it is this book, a blog, an article, or some other high-value project for the day.

Realize that people will treat you the way you allow them to treat to you. It's up to each of us to inform and train others how we wish to be treated.

Here's an example of how your actions set expectations or train a customer, boss, or co-workers. If you answer an email at 7pm on Saturday night, they will always think they can reach you. When I ran my gym, my coaches knew that my phone went on airplane mode from 9pm to 5am the next day.

They also knew that they, along with my parents and my wife, were the only unscheduled phone calls I would answer during the work day. It's important to notify these folks of your intentions and train them to text you first. You can read the text without replying, then decide for yourself if it warrants an immediate return phone call, text, etc.

Even in these situations, *you* are dictating order of importance, not the outside world.

The Power of Saying No

Don't waste your life doing shit you don't enjoy. Being able to say no is real power.

I'm as guilty as anyone for trying to squeeze too many things into each and every day. It's a recipe for stress and burnout.

Less is more.

I think it started out of necessity when I became a business owner/entrepreneur. I literally started with zero clients and no place to train them.

I knew I had to build a client base and quickly build something that resembled a revenue-generating business. If I didn't, I had no income – and that meant I wouldn't be able to feed myself or pay rent.

That was real motivation. This is no man's land, or "the dead zone," as Steven Pressfield calls it. I wouldn't advise this strategy for everyone. You have to know yourself and know how you respond in certain situations. For better or worse, I knew that I needed this type of "sink or swim" environment to launch my dream.

I'm a firm believer that comfort is the enemy of motivation. We've discussed this. Prior to putting myself in this position, I was comfortable enough that I was not motivated to make a change. Removing my comfort (the security of income) was exactly what I needed to spur action.

In those early days, I learned a lot about the difference between entrepreneur mindset and employee mindset. Employee mindset is "do as little work as possible and find as many ways as possible to pass the time from punch in to punch out." Entrepreneur mindset is exactly the opposite: wake up with your feet on fire and get as much done to move your mission forward as possible. Then repeat. And repeat. Usually until burnout or, if sage council is present, until you develop systems that allow you to find balance and protect your energy and time.

Like I said, I was nothing special when I developed this mindset. I just had to – not being able to afford groceries will change a person's motivation level.

I grew my business quickly. That first year, my now wife was doing her first year of residency in another state, two hours away. She kept a small apartment there. We had no TV in our apartment in Roanoke. I literally spent every waking hour building my dream. I learned to strategically schedule every hour from 7am to 9pm with things that would make me better, serve my members, and build the business.

From reading books to improving my weaknesses, to sales calls, to training clients, to building the gym's website and creating marketing content – I became a machine.

Like it does for many entrepreneurs, it became a badge of honor, then a one-way ticket to being stressed out, having not-like-me anxiety.

The solution? I learned many useful strategies that I still use and are outlined here. But once our businesses get beyond survival mode, we must allow ourselves to slow down.

We cannot think, foresee, create, and solve complex problems if we're constantly doing and reacting to incoming stimuli. For long-term success, we need to slow down, plan downtime into our days, and strike our own version of balance.

For me, this was learning to put fewer things on my to-do list and blocking off protected time to do nothing. This space affords us the opportunity to reflect, audit, improve, and identify new ways to continue to grow and evolve. It also protects our precious bandwidth, preserving emotional resiliency.

Remember: our time is most precious resource. How we choose to spend it determines where we go with our life. Take nothing for granted, make no assumptions, and bring your new sense of heightened awareness to the choices you make about how you spend your time.

Goals and Actions, Not Outcomes

Winners focus on winning. Losers focus on winners.
– Unknown

How can we avoid this common mistake? Simple.

I coach my clients and members to focus on actions, not outcomes.

Most people mess up because they focus on outcomes, not the processes that lead to successful outcomes. The problem with outcome-oriented goals is that they tie success and self-judgement to external factors that we cannot control.

If we want success, or *more* success, we should prioritize our actions or how we spend our time.

Every successful person I've coached, interviewed, or studied has prioritized their time and their actions, focusing on what they do each day to move themselves closer to their goal(s).

Stop focusing on the success itself; that's an *outcome*. Instead, focus on your *actions*.

Picasso and Springsteen are two artists so famous, with such large volumes of work that I don't need to give you their first name for you to recognize them.

Picasso is credited with about 100 "masterpieces," a feat that is absolutely ridiculous if you stop to think about it.

Yet, in his lifetime, he created over 50,000 pieces.

Do the math on this. 365 days a year x 50 years = 18,250 days. 50,000 pieces of art in 18,250 days is an average 2.7 pieces per day, if he worked every single day, for 50 years.

And only 100 of those 50,000 pieces are considered "masterpieces" by people not named Pablo Picasso. In case you're wondering, 100/50,000 is less than 1%. Actually, it's 0.2%.

Clearly, he wasn't creating art in the pursuit of critical or financial success. He wasn't doing it for anyone else, or for any reason other than mastery of his craft and to explore the art. Clearly, it was a passion, and he had something to express – either for his own mental health or some other internal drive.

Springsteen lived like a vagabond, played shows for next to nothing, and toiled in obscurity for almost twenty years while he honed his craft. He drove cross-country several times for opportunities to play shows, despite not owning a car, or even having a driver's license!

He can't read sheet music, and in his autobiography *Born to Run*, he tells a story of being given an old guitar with no strings that a friend found in a closet. For six months, he played it on stage in live gigs before a crowd member came up to him and said "that was really smart to string a bass guitar with guitar strings to get that sound." Bruce hadn't known it was a bass until that point.

For six months, he was unknowingly playing live gigs with an improperly strung instrument.

How many times do we let the *how* prevent us from starting? How many times do we let the thought of negative results prevent us from taking action?

How many blog posts have you written? How many songs have you written? Have you played a single gig at your local dive bars or spoken for free at a community event? What are you doing to hone your craft?

Notice how our culture today expects immediate expert status and immediate expert-level compensation without first honing our craft and building toward mastery for mastery's sake.

Forget the outcomes. Focus on actions.

The point here with Picasso and Springsteen is to illustrate the desire of true masters to explore, develop, and innovate their craft.

What if Picasso stopped painting because he wasn't finding validation from the public? What if Springsteen waited until he was an expert with his instruments to start playing publicly? What lies are we telling ourselves about our dreams and passions? What utopia are we waiting for to start making our dreams a reality?

Stop waiting. There is never a right time.

It's Never the Right Time. Do it Now.

Stop waiting for the right time. There is never a perfect or right time.

We can always find a war, a recession, a politician, or any number external circumstances (read: excuses) to delay the pursuit of our passion.

Speaking of excuses, we can have excuses *or* results, but not both.

Procrastination, delay, and avoidance are not the way of the high performer.

Any high performer you can think of would do it anyway, wouldn't they? We've already talked about acting "as if," so it's time to practice that. We must act as if we already are the high-performing version of ourselves. Be that person.

Our fears and the feelings they produce are the only thing standing between us and the highest version of ourselves that we envision. F*ck those feelings.

Stop waiting for confidence or motivation.

Confidence is the decision to try. If we decide not to try or not to start, then self-doubt wins. And the more it wins, the stronger it gets. Having a bias toward action helps us avoid this.

Take action and generate momentum. Use any of the state-altering tools in this book (cold showers, music, dancing, etc.) that you have to in order to overcome inertia. That is precisely why these tools are here.

Get an accountability buddy, make threat bets, post your intentions publicly on social – all have been shown to increase our odds of follow-through.

The most important thing is that we start. Then, keep going. Be you. Not for success, but for happiness. Our happiness depends on our pursuit of our passions. Forget this inner dialogue of, "Should I do it", or, "Do I feel like it?"

If we only did things when we felt like it, we'd never get anything done.

Those slumps that we all inevitably face are perpetuated by the negative feelings associated with inactivity. The fastest, best way to break out of a slump is to alter our physiology with the tools we've discussed throughout this book – a cold shower, Michael Jackson or Prince radio, and a twenty-five-minute timer (Pomodoro) that forces you to get as much done in twenty-five minutes as possible.

This instantly alters our mental state, creates momentum, and begins a new, self-perpetuating cycle of positive motion and productivity before our feelings can talk us out of it. It's all about this activation energy – the push required to overcome the inertia of standing still, or worse, regressing.

Another line from my high school basketball Coach Hart is this one: "We get better or worse every day; we never stay the same."

This sentiment is echoed in Darren Hardy's *Compound Effect*. If we think we're staying the same, we're actually sliding away from our intended goal.

A few other thoughts that help me get moving when I "don't feel like it:"

- The only way out is through.
- Get the day's first victory. Create momentum and win the day. There are no ties. The alternative is to lose this day. I never want to go to bed at night having lost the day.
- Done is better than perfect.
- Focus on progress, not perfection.
- We do the best we can for our current level of development. (But we're always striving to level up that developmental status).
- Do It Now.

We have the responsibility to honor our breath and our instincts with action. Focus on the process. Focus on the things we can control. In a moment of high alignment, define your values, set some clear action items, and begin doing the work.

As you move through your journey, find ways to check back in with your values and goals to maintain alignment and momentum to avoid stagnation or faltering.

The mention of goals leads us to another important

conversation we must have. High performers set goals based on actions and measurable efforts, not hopes. Hope is not a strategy.

For example, if our goal is to get healthy, clean up our diet, and lose some unwanted weight, most people may say, "I want to lose ten pounds."

That's a wish, not a goal.

With some coaching on how to make a SMART goal, that same person may say "I'm going to lose ten pounds in the next two months."

On paper, this is a perfectly executed SMART goal.

S = Specific
M = Measurable
A = Attainable
R = Realistic
T = time-driven

But here's the problem. The goal itself (losing ten pounds) is an outcome. If you lose nine pounds, it's considered a failure. And most people will be disappointed with this result.

Conversely, an *action*-based goal would be to focus on eating healthy meals, avoiding foods that derail progress, and doing the workouts required to reduce unwanted body fat.

These are positive lifestyle changes that can be controlled and measured. As we've learned with health and fitness (and most any other pursuit in life), success is built on consistent lifestyle habits, aka traits.

The action-based SMART goal becomes: "For the

next sixty days, I will eat zero sugar, avoid processed foods, exercise three times a week, and eat five kinds of vegetables every day."

At the end of sixty days, the "judgement" is black and white – there is no grey area. Either you did it, or you didn't. And there is only yourself to blame (or credit) for doing or not doing these actions.

Odds are, a person doing these actions will see a positive outcome at the end of the sixty days, while creating positive lifestyle habits that will stay with them beyond the sixty days.

I use weight loss as an example because it is a common goal and one that I've seen in my nutrition, health, and fitness background thousands of times over the last decade.

The beauty of action-based, SMART goals, is that this template can be applied successfully to any pursuit.

For example, if you want to start a speaking career, set an action-based SMART goal to join your local Toastmasters today, attend their next meeting, and volunteer to speak for free three times in the next month at local events.

Remember: it took Picasso and Springsteen more than sixty days to reach their goals.

It about the process. Mastering the process is about mastering our minds, showing up consistently, even when we don't feel like it, taking ownership and responsibility for our choices and actions.

This is the real measure of health and success. The scale and our bank accounts are bullshit measures of health and success. Many people achieve the numbers they want on both, yet still feel unfulfilled.

Identify your passion, develop it, have a bias toward (and focus on) your actions. Do the right thing for the right people and keep moving forward.

Reflection Questions

1. Your life is perfectly designed for the results you're currently getting. Are you happy with the current results? What would you like to see different? How can you adjust your design to produce the desired results?
2. Where can you say no?
3. How can you maximize the way you spend your time?
4. What can you delegate? Who is on your team?
5. Are your actions congruent with your stated goals and values?
6. Are you happy with your relationship with/ use of social media?
7. Do you set SMART goals? Are you focused on outcomes or actions?
8. Which of the three people in our head are you?

The Three People You Meet in Your Head

Why adults lose that child-like enthusiasm is a topic in itself. For the scope of this book, let's focus on how increased vagal tone (higher HRV) due to increased feelings of safety (defense responses turned off) will enhance our child-like enthusiasm.

Think about the scenarios we explored in the

previous chapter. Kids, whether they realize it or not, feel safe expressing their dreams, no matter how outlandish they are. We even tell them to dream big and that they can be whatever they want to be when they grow up. Adults, on the other hand, do not feel safe expressing similar hopes and dreams – even modest ones, much less the "moonshots" that Peter Diamandis refers to in his book *Bold*. Moonshots are goals that, like shooting for the moon, are out of this world.

Diamandis notes that in order for us to feel safe taking moonshots – or pursuing more modest goals – we must operate from an environment where we feel safe. We must be in environments that celebrate us, not just tolerate us.

Now, think back to our child vs. adult examples. We celebrate, cherish, and encourage our children, even if they're not our own. Adults, not so much. It's no wonder we, as adults, are embarrassed and ashamed to vocalize and pursue our lofty goals.

There is no debating that this negatively impacts our vagal tone, our feelings, and our decisions.

We're all on our own journey. And where we are in those journeys falls somewhere on a spectrum at any given time. In my experience, we spend life in one of three places. I have personified these places, calling them "the three people you meet in your head."

While the place we're currently occupying influences our feelings, and likely our decisions, while we're in that state, it was our previous choices that led us to that place.

As Michael Bernoff explains, our "current

situation is probably a solution to a previous problem." With this perspective, realize that previous problem is now behind us. What now? What would you make your life about?

We must be aware of this and be deliberate about creating the future we want for ourselves.

We need to take ownership for our choices and actions, and in doing so, we also need to realize that decisions we make in any given moment will determine which of these three places we occupy in future moments.

Remember: your life is perfectly designed for the results you're currently getting. So if you're where you want to be, keep doing the things that got you here.

If you're in a place that you don't particularly care for, know that the choices you make in this moment can elevate you and move you closer to the life you want to live.

The three people we can be, or the three mental spaces we can occupy, are:

1. Joy, in Doing: of doing or having done;
2. Frustrated, in Purgatory: anguish and frustration of not doing/trying/or pursuing;
3. Embarrassed, by Failure: embarrassment of trying and "failing."

Person One: Joy, in Doing

Of the three people in our heads, we want to spend as much of our time as possible being Joy, in Doing.

Happiness is often defined as the opportunity to pursue our calling. That's it — figure out what captures your imagination and build a life around mastering it and helping others do the same.

It sounds too simple to be true. Why do we make it so complicated? Why do so many of us fall short of this?

To paraphrase some Eastern philosophy, we are entitled only to our labors, not necessarily the fruits of them.

In today's society, especially in the Western world, we are in the results business. While I realize results drive capitalism, relying on results to be the determination of success (or happiness) puts happiness outside of our locus of control.

Locus of control is an intriguing area of study dating back to the 1960s and a man named Julian Rotter. In short, people with "an internal locus of control believe that they are responsible for their own success. Those with an external locus of control believe that external forces, like luck [or fate or other people] determine their outcomes."

As you might infer, the more one's locus of control tends to favor "external," the more fear, luck, anxiety, and fate dictate their behaviors and happiness.

We don't want to live this way. These are feelings and emotions tied to negative physiology, decreased vagal tone, reduced emotional resiliency, and constipation (emotional/energetic blockage, as well as digestive).

On the other end of that spectrum, a person with a strong internal locus of control eschews fate, fear, and luck in favor of ownership over their own action, hard work, and decisions.

The joy of doing is *the process*, if you will.

As the saying goes, and the now popular Ryan Holiday book is titled, "the obstacle is the way."

Remember this book is called *F*ck Your Feelings*, so we do want peace, love, satisfaction, and fulfillment, but those things are the end results, not the actual pursuits.

And we don't need scientific equations to find them. We simply need to listen to the voice deep within us and honor that calling as it is our truest self – our authentic self and our reason for being on this earth.

Did I just say simple? Yes, I did. Easy? No. If it were easy, everyone would be doing it.

Simple merely implies that something is not complicated. Easy implies a lack of effort. Pursuing your true calling and living your authentic life will certainly require a great deal of effort, discomfort, and mental fortitude.

If you need help figuring out your true calling, you're going to have to meditate on that one for a while. If you're like most people, you already know what it is – that business idea you always had, or the book you've always wanted to write.

I don't think we "find" our passions. We already know what our values and interests are. We need to develop them. Time, repetitions, and deep knowledge in a field turn interests into passions and help us move toward mastery.

This is the path.

You can always use the question from the movie

Office Space: "If money wasn't a concern, how would you spend your days?"

What would bring you excitement day after day and make you climb out of bed with purpose for years on end?

For me, it's making the world a stronger, healthier, happier place.

When we find and commit to following that purpose or calling, the universe conspires to help us. It's an odd but powerful phenomenon.

We may not know *how* to make it happen. I'd say 99% of people who start on these journeys have no idea how to make it a reality. And each of these journeys, no matter how clear the path may have seemed, will always be full of roadblocks, twist, turns, obstacles, and setback.

We've all seen the image showing how success is not a straight line:

Success

what people think it looks like

Success

what it really looks like

I'm not saying it will be easy.

And I don't want to say "it's worth it" because that sounds like a cheesy calendar quote.

So I'll say it another way: F*ck Your Feelings of being scared, intimidated, or insecure. We get one life, and tomorrow is not guaranteed. To delay our pursuit, to delay action is to take time (life) for granted.

We spend 50% of our lives working and another 33% of our lives sleeping. That leaves about 17% of our life to shower, commute, watch TV, and get sucked into social media, *or* we can get out of the matrix, take control of our life, and spend our precious time with potent intent in the pursuit of the *real* reason we were put on this beautiful earth.

Why not spend your life doing the thing that brings you the most joy and happiness?

To spend it any other way is to dishonor your soul and waste this precious gift of life. These are the other two people in our heads, filled with pain and frustration, and as you'll see, they're not the versions of ourselves that we want to be.

When we set out on this journey, we create positive momentum that transfers to every aspect of our life. Amazing, unforeseen things will happen that enrich our lives and buoy our pursuits. As Paulo Coelho says in his bestselling book *The Alchemist*, "The world conspires to help you."

We meet other people on their own journey who become friends, mentors, and peers. We develop a self-confidence we never imagined we could possess. Our growing network of friends provides us support and new and bigger opportunities. Relationship capital is the most important capital we'll ever possess.

The lives we positively impact and the new knowledge and skills we develop make us more useful and valuable human beings. This bias toward action creates momentum and confidence. It's a self-perpetuating cycle. We simply have to choose to begin, to generate the inertia.

And like Newton's first law of motion, an object in motion stays in motion unless acted upon by an outside force.

But – and here's where we all have a tendency to get in our own way – if this action is the pursuit of a true calling, there is an internal force that can slow, stop, or completely derail our forward motion.

Steven Pressfield calls this force "resistance." Mostly, resistance is fear. It's that lower version of ourselves. It's our ego that wants to avoid pain. It's our lizard brain, seeking tribal acceptance and fearing what may happen if we refuse to play small and strike out to see what may be if we pursue that which excites us. It's our feelings of fear, self-doubt, and that destructive narrative in our head about who we tell ourselves that we are not.

Persons two and three in our heads are these lowers versions of ourselves. The versions that lose the battle to resistance. The versions too scared to step into our greatness and be who we are truly capable of being.

F*ck them. Who are they to stop us from living our true purpose?

Person Two: Frustrated in Purgatory

This is the real hell.

Most people think failure and the embarrassment they assume will follow it is the worst, but the purgatory of never trying is the real hell. This is the real "no man's land" that we should seek to avoid.

Trying and failing is not as bad as we think, but we'll cover that in a bit. For now, let's focus on the anguish and frustration caused by letting that fear or doubt paralyze us.

No doubt, we've all heard the saying, "Fear kills more dreams than failure ever does." It's a great quote and has motivated me on many occasions, but it's not entirely accurate. It's not fear; it's anxiety.

Semantics? Maybe. But maybe not, if we look at fear and anxiety from a neuroendocrine perspective.

"Scientists generally define fear as a negative emotional state triggered by the presence of a stimulus (a snake on the trail in the woods) that has the potential to cause harm, while anxiety is a negative emotional state in which the threat is *not* present but anticipated.

We sometimes confuse the two: when someone says he is afraid he will fail an exam or get caught stealing or cheating, he should, by the definitions above, be saying he is anxious instead." [261]

Fear is what we experience in the presence of a threat. Anxiety is what we experience when we anticipate a threat, but the threat is not actually present. Learning this nuanced difference can be a game-changer when it comes to mastering our minds.

Like stress, some anxiety is good. But the dose makes the toxin. And like stress, chronic (or pathological) anxiety is not beneficial. As fear researcher and behavioralist Dr. Thierry Steimer writes, "Pathological anxiety interferes with the ability to cope successfully with life challenges." [262]

That's not the only evidence that fear and anxiety negatively impact our behaviors. A 2004 review of the consequences of fear concluded that our misplaced fears (or anxiety) are, in most cases, more harmful than the thing we fret over.

In their words, "The hazards of risk misperception may be more significant than any of the individual risks about which we fret." [263]

This anxiety negatively impacts our vagal tone in the moment, drives negative feelings and impairs our decision-making. What's worse is that this anxiety, if left unchecked, can actually alter our traits as individuals and impact offspring through epigenetic shifts.

In an example from the animal kingdom, "Predator exposure to wild prey animals has been shown to lead to 40% less offspring production and it is linked to glucocorticoid elevation in the parents. Multi-generational stress has been demonstrated in snowshoe hares which may increase an adaptive predator response in future offspring." [264]

Think about rabbits, some of nature's most prolific reproducers. They produce 40% less offspring because they encountered a predator and now fear that they *might* be eaten.

If anything, facing mortality and the reduced chance of offspring survival, they should *increase* reproduction output to improve chances of species survival. But they don't.

Their fear (err, anxiety) over a potential threat changes their individual behavior (the perception of threat activates defense response, which decreases vagal tone, and alters decision-making, and of course, less sex) and threatens the survival of their species with a 40% reduction in offspring and altered genes (epigenetic expression of anxiety) in subsequent generations.

That's a big freaking deal, and we're foolish if we think this occurs only in rabbits. Human behavior is equally shaped by past experiences and the anxiety we experience if we don't properly deal with those experiences (big or small).

The same study continues, "The fear neural circuitry includes; amygdala output circuits that directly activate the sympathetic nervous system [fight or flight] and also the hypothalamic pituitary adrenal (HPA) axis, thereby including stress hormones in the negative emotional response." [265]

This is more evidence that directly supports what we've previously discussed – the importance of tribes/communities that make us feel safe (shuts down defense response) and the subsequent high vagal tone for increased emotional resiliency that facilitates heightened awareness.

Elon Musk has famously been quoted as saying we should aspire to increase human consciousness and strive for greater collective enlightenment. Well, this is how we can achieve it.

It's also more reason to actively seek more time in the parasympathetic state with activities like yoga, meditation, saunas, and breathe work, as we've discussed throughout this book. Remember: higher HRV also reduces magnitude of response to future fear and anxiety-inducing situations. [266]

Inaction, and failing to honor the inner voice, gives rise to haters. This purgatory of inaction is also the space that gives birth to "haters."

This land of frustration and unfulfilled lives is where haters live. As world-renowned strength coach James "Smitty" Smith once told me that haters are simply unrealized potential.

They don't really hate others. Their hate or frustration is with themselves, and it is directed outward toward others. They are frustrated by their own lack of accomplishment and the voids in their own souls, so they direct this outward at others.

Think about it. Have you ever known a genuinely happy person to be a hater? Doubtful. If they exist, they're the exception, not the norm. Similarly, have you noticed that very few haters are more successful than the people they hate on? People who are doers and focused on their own selves and their actions don't have the time or desire to be haters.

Bored, frustrated, miserable people throw rocks at shiny things. It's easier to destroy and bring down than to create. It's classic projection and lack of personal responsibility – the psychological method of protecting the ego (feelings) by denying the existence of something (or lack of something) in one's self while attributing it to another (or others).

In most cases, the people doing this are not equipped with the emotional maturity to realize what they're doing.

We see it in the form of spousal sabotage all the time in health and fitness endeavors. Let's say a wife wants to lose fifteen pounds, eat healthy, and "get fit." She starts eating better, going to the gym, and making changes and better choices. She's eating differently, buying different foods, and cooking different meals. She has stopped indulging in many past behaviors.

The husband is now forced to admit to himself that he isn't taking action, but he doesn't know how to deal with those uncomfortable feelings so he lashes out, taking those feelings out on the person who "makes him feel that way" – his wife.

This sabotage may be conscious or subconscious, ranging from deliberately eating pizza or candy in front of her, to hurtful comments, distancing himself in the relationship – any number of things.

The same can happen outside the fitness realm – it happens anytime we grow, evolve, and move forward. We need to be aware of this on both sides, as these efforts say to the people around us that our current situation is not ideal.

If you find yourself being a hater or saboteur, *stop*. Realize why, identify the triggers, and meditate on what about those triggers is stirring the feelings in you. Then stop being a dick, and start focusing on your own shit.

We must take responsibility for our own growth and our own actions. We should always handle our own shit before looking at other people's shit.

Since you're the type of person invested in reading this book, this is hopefully more of a reminder than a revelation.

Use the strategies and tools in this book to move

from this purgatory to the person you want to be – person one.

If you find yourself dealing with haters, realize that you're doing something they wish they could. You may (or may not) be succeeding at the level you want, but *something* you're doing has grabbed their attention and made them experience feelings, thoughts, and emotions that they are not dealing with properly.

That's on the hater, not on you. Remember that "haters are never doing better than you."

On the other hand, successful, happy people only want to see more of that in the world. They're likely to be your biggest supporters and cheerleaders. Seek them out, surround yourself with this kind of person, and leave the haters behind. (If the hater is family, a friend, or a loved one, explain the situation to them and give them a chance to get on board, but make it clear to them that your train is moving forward – it's up to them to get on board or get left behind).

Coming up, we'll talk about how to surround yourself with positive, driven people who hold you to a higher standard and help you grow and develop as a human.

Person Three: Embarrassment of Failure

Worse than any wreck is knowing that's in you and you never let it out. – Drive By Truckers

Embarrassment of Failure, or person three, is largely imaginary. Anyone who has ever tried and "failed" will tell you this space is far less scary in reality than it is in our minds. We fear something that really isn't that bad.

We think our failures or missteps will be broadcast, mocked, and turned into Internet memes, as if our lives hold the public's attention like Donald Trump or Kim Kardashian.

Our egos would be in heaven if that were true! We can only wish!

Truth is, nobody f*cking cares. Very few people will even notice your "failure." Maybe 300 people in your town? 1,000? 2,000 Facebook "friends," of whom only 10% see your posts and only 100 are people you care enough about to wish happy birthday?

Here's the reality: we are the only ones who view ourselves in a negative light after a swing and miss. (Provided our attempts are legal, moral, and well-intentioned). And the best part about this is that we can control the narrative we tell ourselves about ourselves.

Of those very few who are even aware of this "failure," most will actually hold you in higher regard because you actually attempted something.

The doers in this world respect other doers because they realize what is required to even try. And the non-doers will put you on a pedestal of sorts in their minds because you're likely to be one of the few friends, peers, or family members in their life who actually had the courage to try – and this courage, ill-fated or not, will inspire them. (Whether they take action or not is another story).

You're different. You dared to escape the drudgery of "normal." You exited the matrix, left the safety of the tribe's fire-lit circle and explored the boundaries of possibility.

In your mind as well, you'll become a doer. The first attempt is usually the scariest. And like a baby learning to walk, or a child learning to ride a bike, it just gets easier. Our attempts get stronger, and we go further and further each time.

Learning first-hand that you can try and walk away relatively unscathed, even stronger, in most instances is the most empowering thing you can ever do for yourself.

If you did it once, you can do it again.

In fact, approach it like the kids we've discussed previously – what if a kid fell the first time trying to walk and then said, "Well, f*ck it, I can't walk, and I guess never will."

It's mind-boggling how often adults take this very approach to the pursuits that are supposed to be their *dreams*!

Thomas Edison famously tried nearly 10,000 different ways to make a light bulb before he succeeded. How many of us are willing to try even 100 iterations of something before we give up and write ourselves off as failure?

I have not failed, I just found 10,000 ways that did not work. – Thomas Edison

"Fail forward," as the saying goes. Learn from your missteps. Don't consider them failures. Consider them iterations, like Edison and the light bulb.

Remember our story of Springsteen playing low-paying gigs on the Jersey Shore night after night. Was that failure? Or was that the process of become great?

Likewise, you're not a failure if you write a blog with only ten readers or speak to a crowd of three, or write a book that sells 100 copies.

The "failure" is only in your head, because you're comparing what is to what you think should be. This is the comparison inferiority complex we mentioned previously. This is focusing on outcomes, not actions. This is "losers focusing on winners" instead of focusing on actions that lead to winning.

It's all part of this journey we call life. It's about action. Don't be a spectator. Don't let someone else control or dictate your path.

To paraphrase Teddy Roosevelt's 1910 speech "Citizenship in a Republic," it is not the critic who counts, but the man bloodied, dirty and dusty, standing triumphantly on the arena floor. Not the person in the stands or on the sideline, but the person on the field and playing the game with everything they have.

Live your life. Get off the sidelines. Contribute. Get involved. Honor the life that your parents brought into this world.

If we're lucky, we'll live to be old men and women who look back on our lives — at which point the question will be: as we sit in that rocking chair, will we be content with the path we chose for our life?

When the distractions are peeled back — no more social media, no more Netflix — will we be satisfied with our life's work? Did we serve our purpose? Did we answer our calling? Did we live? Did we love? Did we matter? What legacy will we leave behind? What contribution did we make to society?

As it is so ego-shatteringly framed by Darren Hardy in *The Compound Effect*, we are writing our own eulogies every day through our actions (or inaction).

You're writing the story of your life. What will you write? What legacy will you leave? How would the ultimate version of you move through life? Act As If – act as if you are that person.

Activities

I have a FREE three-video series that walks you through the following activities that I use with all my consulting clients:

1. Define Your Values;
2. Act As If (The Jedi-mindset cell phone trick);
3. Using Choice Architecture to Create Better Defaults.

You can access this video series for at: ryanmunsey.com/define-your-values.

Smile

Typically, we associate a smiling face with a person who is happy and in a joyful mood.

But the smile itself isn't a feeling or an emotion. It's a physiological occurrence, and as we've learned, it's our physiology that drives emotions, feelings, and thoughts.

So the smile isn't just the result of a positive mood; it can also be the cause.

As Darwin wrote in 1872, "The free expression by outward signs of an emotion intensifies it."

A 2009 study from the University of Cardiff in Wales found that women given Botox to prevent frowning experienced greater happiness and less anxiety.

Study co-author Michael Lewis noted that "it would appear that the way we feel emotions isn't just restricted to our brain--there are parts of our bodies that help and reinforce the feelings we're having. It's like a feedback loop."

A follow-up study in Germany supported these findings when it used functional MRI to scan the brains of subjects asked to mimic angry faces. Those injected with Botox and inhibited from frowning displayed reduced activity in their amygdalae, hypothalamus, and parts of their brain stems, confirming that the facial expressions – at least in part – drove brain activity related to feelings and emotions. A 2008 study published in *The Journal of Pain* found that people who frown during unpleasant procedure experienced more pain. [267]

These findings give scientific validity to the phrase "fake it 'til you make it." If we're not feeling it, put on a smile and trick your brain into thinking you're happy.

This is one reason I love dancing to Prince, Footloose, or Michael Jackson. It's almost impossible to give those activities our full effort and have transform our physiology.

When I was opening and building my performance-training facility, House of Strength, I was taught to always make calls and answer the phone with a smile because people on the other end could tell a difference.

I immediately noticed a difference in the way I interacted on those calls when smiling, and in turn, I enjoyed much more desirable outcomes from those phone calls.

Always smile on the phone and smile more in life – for yourself, your health, and your happiness.

Smile. Repeat after me: "I'm f*cking awesome."

Keep repeating this until you believe it or start smiling about the ridiculousness of this activity. Either way, physiology is altered. Now, show up. Do the work. Repeat as often as possible. Enjoy the journey.

CHAPTER 9:
Habits of High Performers

Success Leaves Footprints

I wholeheartedly prescribe to this adage.

Rather than look for differences, I look for the shared ground between proven systems and successful people.

Whether we're examining diets, business strategies, fitness modalities, or high performers, the qualities and aspects shared by them are the traits on which we should focus.

For example, bodybuilding, CrossFit, Olympic weightlifting, Strongman, and powerlifting are all unique disciplines within the fitness community. On the Internet, there is much in-fighting between these groups about which is best. The discussions often focus on the differences. I prefer to look at the overlapping values they all share.

For example, each discipline utilizes a few similar movements – pressing, pulling, and squatting. Each understands the value of repetitions, accumulating adequate volume and increasing strength over time.

Taking this high-level view reveals a basic truth that anyone can apply to their fitness pursuits for success: focus on a few basic lifts, get really good at them, and get stronger over time. The rest is splitting hairs.

Let's apply this concept to the habits of some of the high performers I've worked with, who in turn have influenced me and much of this book.

Habits of Successful People

I've been fortunate to work with a handful of Olympians in an effort to deconstruct their habits and find the common traits they all share. What makes the Olympians stand out in my mind is the sheer magnitude and length of their preparatory cycles. The Games only occur every four years!

In a society where most people struggle to maintain consistency for a month, ninety days, let alone an entire year, this demographic signs up for a minimum four-year commitment. As a matter of fact, each of the athletes I've met and spoken with for this book competed in multiple Games, which means their training and preparation for a single event spanned the better part of a decade, from eight to twelve years. We could even say they've been preparing for this over a lifetime, if you account for their training and development before the Olympics actually became a reality.

This unique group of high performers shares a few common qualities that I believe are prerequisites for success. What makes these traits and practices even more compelling is that these athletes come from four different countries (US, UK, Canada, and the Netherlands), have different socioeconomic backgrounds, and competed in different sports (track and field, boxing, rowing, and beach volleyball), making their shared traits independent of sport, culture, or background and demonstrating their application across domains.

In the list below, I've replaced "Olympians" with "people," since I believe these traits are applicable to success in life, not just the Olympic Games.

1. Successful people are lifetime learners. I have yet to meet a successful person (Olympian or otherwise) without an insatiable appetite for knowledge and a relentless quest to hone their craft and get better as human.
2. Successful people have an extraordinary belief in themselves. I use the word extraordinary purposefully here, meaning beyond that which the normal person possesses. Maybe it's the countless early morning wake up calls or the thousands of reps, but I think it's their knowledge of their own resolute commitment to a purpose and the fact that they have demonstrated (to themselves) their commitment to excellence despite any obstacle that fosters this belief. Their daily actions and consistency are what is truly extraordinary. Action creates momentum, and momentum creates confidence. Get started, be consistent, keep moving forward, and build extraordinary belief in yourself.
3. Successful people have taken accountability for their choices and decisions. Read the book *Extreme Ownership* by Jocko Willinck and Leif Babin. Things don't happen *to* us. Remove the victim mindset. Everything in this world is vibrations, and nothing can happen to us vibrationally that we're not a party to. Take

responsibility for your choices, actions and decisions. The beautiful thing about embracing and embodying this mindset is that it puts us 100% in control. The moment we make this choice is the moment that everything changes. The story of your life is being written – make damn sure you're the one holding the pen.

4. Successful people have the ability to focus on actions required to move forward. Olympians, SEALs, and CEOs all focus on the immediate tasks in front of them and the things that they can control – their actions. SEALs have a thought process of "just make it to the next meal" that gets them through their brutal training. There is famous saying about eating an elephant: it must be done one bite at a time. Whether our goal is four months, four years, or forty years, we must break it down into bite-sized chunks and focus on the daily tasks that must be accomplished to close the gap from where we are now to where we want to be.

5. Successful people have the ability to ignore people outside their camp. There will always be people who don't believe in us, who want to see us get hurt if we fail, or don't want to see us succeed. None of those people have a place on our squad or in our community. Successful people know this; eliminate them and surround yourself with people who are part of the mission or, at the very least, believe in it.

<u>Mindset Secrets of Navy SEALs (Hint: they love</u>

choice architecture)

One of the Navy SEAL operators I have been fortunate to befriend is a man who spent a few years as a BUDS instructor. We'll call him Ryan. BUDS, if you don't remember, is regarded as the toughest military training in the world, with an 80% wash out, or failure rate.

Known for having one of the highest washout rates among all BUDS instructors, I asked Ryan if he was actively trying to get candidates to quit. His reply surprised me. He said no. He wasn't focused on making people quit; rather, he was trying to identify the people who possessed the qualities required to be an appropriate foxhole partner and eliminate (as quickly as possible) those who did not possess those qualities.

I also asked Ryan about his own experiences in BUDS and how he made it through. I absolutely love his strategy. As we detailed in the Choice Architecture section of this book, he limited his potential choices. Where many candidates go in wondering if they'll succeed, Ryan gave himself two options, neither of which involved quitting: succeed or die trying.

Sure, this is an extreme case, but it shows an obvious example of intentionally, consciously restructuring our realities and creating frameworks and default choices that will enable us to succeed rather than open the door for failure.

Not once was quitting something that entered Ryan's mind when he was a BUDS candidate. There were several occasions he thought he might die, and he told himself to keep going, and that if he passed out or died, it was worth it – that he wasn't leaving a quitter.

Aside from limiting potential choices, Ryan also stressed the importance of stress inoculation.

We'll see this again from other successful people through the practice of repetitions, both physical and mental (visualization, etc.), but realize this is also a real-life example of getting comfortable expanding our comfort zone.

Think of it as opportunity to train your brain and body to avoid sympathetic freak out in times of duress through inoculation via many various repetitions. The more we do this, the more we learn to relax and perform from a place of safety and comfort (increased vagal tone), rather than constipated fight or flight.

Not everything Ryan shared with me can be shared publicly. The following bullet points are my actual notes from my conversation with Ryan for this book:

- Motivation has to be intrinsic. Is what's inside us stronger than what's outside?
- Pain is the gift that nobody wants.
- The problem is you don't know what the problem is.
- Utilize hate and ego when necessary (realize this is a short-term solution).
- How you do anything is how you do everything – it's all about discipline and consistency.
- Get used to adversity – it creates resiliency.
- Ryan refuses to go to bed each day without making himself face adversity every day.
- Your belief system – know your *why*.
- Remove quitting as an option (limit potential

choices).
- Make mini goals and reward yourself for mini goals to build and keep momentum (making through another day).
- Is it really about resiliency? Or is about the purpose/passion, if connecting?
- Find your purpose.
- Use affirmations – morning and night-

<u>The Everest Summiteer</u>

Another successful human who lent insight to this study is Maria Granberg, a mountaineer who, at the time of our conversation, was the most recent person to stand at the top of the world.

She is an amazing athlete who spent as much time (if not more) training her mind for her grueling expedition as she did preparing her body.

She knew that her mind would present her most formidable foe during her climb, so she practiced neurofeedback, meditation, gratitude, and other awareness and consciousness-enhancing techniques to make it a strength that help propel her to success.

Success cannot be something you define by external factors. As Ryan the SEAL eluded to, most truly successful people have an intrinsic definition of success, not an external measure.

The moment we allow our level of success to be determined by other people's opinion is the moment, we lose control of our own happiness, self-worth, and direction in life. If we do this, we'll forever be chasing an ever-moving horizon seeking the validation, love, or admiration of others.

Very often, a large number of those who make up the critical masses that are "required for fame" don't even know you personally; why spend your time trying to earn their stamp of approval?

To paraphrase Lao Tzu's *Tao Te Ching*, this is not the way, if we are to be satisfied in our life.

We need only to validate that inner voice within our souls – our true selves. The self, not the ego.

This is where true happiness, success, and contentment stem from. Shedding the ego, pursuing the calling of the self (soul).

If our (your) success is determined by extrinsic factors that you cannot control, then you've given away your power over your own happiness. You will forever be powerless in your pursuit of it.

The whole point of this book is to empower you to take control and direct your life in the way you want to go; to honor your soul and pursue that fire that burns within you.

Success, however you define it, has to have a component of intrinsic measurement (I'm proud to say that I wrote this section before talking to Ryan).

Are you following your authentic path?

Most of the high performers I've been around share one massive commonality: they're doing what they love. They're pursuing the thing that makes their soul come alive. And because of their passion, they find that they have to hold themselves back and find balance more than they need that kick in the ass.

I don't think this state of motivation or drive is a quality that eludes others. I think it's a quality that develops when we find and begin to focus on the thing we're put here to do.

For most, when we get on that path, we find peace, fulfillment, and a self-perpetuating cycle of drive and motivation to continue the pursuit.

For these reasons, I urge you to pursue happiness, peace, fulfillment, and contentment rather than chasing success. You're likely to find these are the worthiest of pursuits.

If you need more convincing, recall the classic quote from Paulo Coelho in *The Alchemist*:

When you pursue your true calling, your authentic purpose, "all the universe conspires in helping you achieve it."

Science of Accountability

Speaking of getting a boost from co-conspirators, research shows that accountability actually influences our decisions to a greater extent than money or facts about the decision.

A 2014 meta-analysis on emotions and decision-making shows that there are at least five studies demonstrating that the promise of financial gain or saturating the decision-maker with cognitive facts about that domain do not move the decision-maker to make logic-based decisions over emotion-based decisions. [268]

However, that same analysis documented that when held accountable for their decisions, people actually make better, more logic-based decisions compared to feelings or emotion-based decisions.

In other words, the knowledge that their actions would be judged by peers drove better decisions to a greater extent than financial benefit or actual data around the decision itself.

This is further evidence to support our biological wiring to be part of the tribe, as it shows that we would rather be viewed in a high regard by the tribe than make a decision based on how it would benefit us financially or how the facts say we should decide.

If we go back to our high performers, the Olympians, the SEALs, and the Mt. Everest summiteer, we'll see that each was in situation where they were accountable for their actions — either to themselves (judged against their own standards/values) or their coaches or teammates.

Their pursuits and their daily actions impacted others, be it their teams, their coaches, their fans, etc. We can use this information to find ways to hold ourselves accountable for our daily actions.

I work with a lot of people on their nutrition habits, so let's look at cheat meals, a common habit of many people. Let's say you're one of the many folks who drops an extra $100 every weekend on a re-feed, splurge, vice, binge, or cheat day (whatever the name is).

This isn't about the efficacy of "cheat meals" — stay focused on the message, not the example. To paraphrase the Buddhists or Taoists, don't mistake the pointing finger for the moon.

Many people want to stop doing this but struggle to do so. The data-based decision is to look at the money spent and potentially saved.

That $100 a week, over a month adds up to $400, which could be spent on a car payment. Literally, a new car's worth of money being spent on "splurge food" every month. No matter how impactful that logic is, most people, in the moment, would opt for the food. I've seen it time and time again with clients and been there myself.

This is precisely why awareness, limbic system control, and being able to consciously change our physiology in the moment is so crucial in order to improve decision-making to maintain forward momentum.

What's fascinating is that neither of those logic-based thought processes will have as much benefit for driving behavior change as having to self-report choices to a group of peers. A trainer, nutritionist, or accountability buddy/group will be a greater deterrent in the moment that either of the first two, seemingly more logical decision-making processes.

This speaks to the power of emotions and feelings, that despite overwhelming data along with the promise of financial gain, we still opt for that emotional release or immediate fulfillment. We know it's case of temporary, short-term gain that leads to long-term loss, yet we do it anyway – time and time again, according to the research.

This is where it gets interesting. The research shows us that a better deterrent for those actions is accountability, i.e., having a group or community to which you have to report your decisions or your actions. Holding ourselves to an agreed-up tribal or communal standard will drive better decisions to a greater extent than having the data or the potential upside of financial return.

Another study actually found that simply telling someone about your goals increases your odds of achieving the goal by 65%, while creating an accountability system increases your odds by 95%. [269]

Keep this in mind the next time you just want to curl up in your sweatpants and eat donuts, cookies, cake, or ice cream and drink wine while you binge-watch Netflix in an effort to just turn off your brain.

You're not alone in seeking that vagal tone comforting activity. Everyone will encounter this barrier, but how we deal with it is up to us.

Visualization and Mental Reps

The Olympians I talked with all used various methods of visualization and meditation as well. From neurofeedback, to positive self-talk, to the removal of certain choices, the high performers purposefully, intentionally prepared the space between their ears to enable their future success.

When they finally got their moment, they had been there hundreds, if not thousands of times before in their minds. They were prepared, and they succeeded because they were able to go on autopilot in the moment – and that autopilot had been programmed over their years of training to be the exact scenario they wanted.

For successful people, these are habits that are practiced daily. It's part of the routine and they do it no matter how they feel.

What internal dialogue are we having? What mental reps are taking?

Do you struggle with making it to the gym? What do your mental reps look like? These don't even have to be actual reps in the gym, I'm talking about the days, nights, or lunch breaks where you planned your workouts. Do you see your future self getting there or are you showing yourself images of failure?

If you struggle with binge eating or falling off track on your diet, what do your mental reps look like when you think about events that trigger these detours? Do they show you succeeding in the future or do you focus on failing in the past?

If you struggle with binge eating or falling of track on your diet, what do your mental reps look like when you think about events that trigger these detours? Do they show you succeeding in the future or do you focus on failing in the past?

Beware the negative mental reps. We've all heard how powerful visualization can be for creating success in the future. But perhaps more important when trying to develop confidence and discipline and master our minds is to be aware of what I call "negative mental reps."

Negative mental reps are the scenes where we envision ourselves skipping workouts, binge eating, or procrastinating instead of writing our book.

Many times, we play out those scenes in our minds because that's what we've done in the past. Other times, we actually romanticize these distractions because, simply put, they're more fun in the moment than the hard work that our life goals will require if we are to make them a reality.

Again, our ability to maintain a 50,000-foot,

objective view of our minds allows us to be aware of these thoughts. As it is this awareness that affords us the choice to continue playing these negative mental scenes or the choice to see ourselves choosing actions that align with our values and goals.

We must realize and respect the power that negative mental reps have to bring those visions into reality. The more we do this, the more this becomes our reality. Remember our discussions of habits and actions and how they're nothing more than neurological wiring. With negative mental reps, we're literally wiring failure into our life. Those negative visualizations strengthen that pathway, creating a highway where we want an unused dirt road and literally reinforce the undesirable behaviors.

For this reason, avoiding negative mental reps may be more powerful than utilizing positive visualization. Either way, you're going to imagine the future – you might as well do it in a way that cultivates success.

Beware your negative mental reps.

I promise you, the scripts that we write in our heads will be the scenarios that play out when we live out these moments. Look no further than our high performers for proof of this.

All of these are methods of controlling neural activity and physiology during training and before the actual performance to ensure the brain does the right thing when it matters most.

Remove the Ego

You might think SEALs, Olympians, and people who climb Mt. Everest define themselves by those achievements.

They don't. This is another important insight into the secrets of their success. The most successful people are able to remove their ego and separate themselves from their accomplishments.

As another Navy SEAL once told me, "I'm not who I am because I was a SEAL. I was a SEAL because of who I am."

This reminds me of a warning from Tom Bilyeu, "What we build our self-esteem around matters."

In context, Bilyeu was commenting on our propensity to value easily measured standards like income or body fat percentages in lieu of more important qualities like skills, education, compassion, accountability, and other qualities that directly impact happiness and success.

The larger the mission, the bigger the goal, the more important it is to maintain an objective, analytical view of what the mission requires, as well as to be able to assess what our own strengths and weaknesses are. This step is impossible if we're living in and acting through the ego.

Our high performers only reached their summits (metaphorical and literal) because of their ability to operate independently of their ego. They maintained objective, 50,000-foot views of their missions and their actions, and they constantly worked to turn weaknesses into strengths so that they would be prepared when their moment to perform came to fruition.

This objective view also afforded them the ability to recognize the difference between preparation time and performance time. BUDS, Olympic training camp, practice climbs, and workouts were all dry runs designed to simulate or prepare them for the real moment. They were not the main event, and successful people rarely confuse the two.

They may have down days during training, but they realize it's part of the process. By not entangling their ego with their pursuit, they're able to prevent those single days or a single mood from interfering with their long-term plans. They continue showing up, doing what they're supposed to do, without letting it define them.

Ultimately, when it comes time to shine, they're ready. Performance day is never their first "reps." They've done it so many times before, that the neural "path of least resistance" is exactly the one they've programmed.

Can we say that about anything in our life? Have we put in the preparation necessary to be successful? Have we prepared our minds, taken the objective view, separated our ego from our pursuits, performed the reps (both mentally and physically) to be able to expect success?

Successful people are the ones who can answer yes to these questions. The inability to answer yes to these questions is a clear indicator of the source or cause of self-sabotage and/or lack of success.

This brings us back to our conversations about awareness. Awareness creates choice.

To master the mind (or self) is said to be true power. While this may be true, I'll argue that we cannot control – much less master – what we don't understand.

This book has explored our neurobiology and

given us and intimate understanding of what the brain and body are biologically wired to do.

This understanding and knowledge afford us the awareness that creates the ability to choose our course instead of being along for the ride.

The strategies presented from personal experience, clients and case studies of high performance individuals have armed you with a multitude of tools that are now at your disposal to slay the dragons between your ears and live your life as the highest expression of yourself.

It doesn't matter how you do it – whether we use meditation, neurofeedback, Holosync, yoga, or cold exposure – what matters is that we need to learn to control our mind. All of our high performers exhibit extraordinary control over their mind. It's all about the evolved prefrontal cortex being in control over the primitive limbic system and keeping vagal tone in a range conducive to high emotional resiliency.

Decide that you're going to do something, commit, and give it 100% total human effort. F*ck Your Feelings along the way. Do it anyway. Do it whether you feel like it or not. Take it one step at a time, and focus on progress, not perfection. Keep moving the chains. Make shit happen and do it with a community of support that holds you accountable and challenges you to be the best possible version of yourself that you can be. Be responsible for your own actions.

Think about any superhero or movie character that we hold in high regard – they have a code that they defend and they overcome adversity.

Adversity is going to happen. That is inevitable. What do you do when it happens?

The better trained we are in the art of awareness, consciousness, and prefrontal cortex over limbic system, the better our ability to recognize, diagnose, and course-correct. When we have not developed this skillset, we're susceptible to the negative moods, poor decisions, and downward spirals associated with adversity.

Adversity (think momentum in a football game) is nothing more than energy. We can always ask ourselves, how are things vibrating at this moment and where is our energy going? In our best efforts to be anti-fragile, we need to resist the negative downward spiral, be solid, and stand our ground – then, using our awareness, choose to move toward our goals instead of being pushed passively away from them.

Without adversity, there is no triumph. Expect it. Prepare for it. Then perform and prevail.

Heroes have a code. They have values they stand for and defend no matter what. Decide what it is that you believe in and stand for, then hold the f*cking standard and f*ck those pesky feelings that try to derail you along the way.

What is your code? What do you stand for? What positive impact will you make? What will your legacy be?

Reflection Questions

Before we close this book, let's reflect on why we're embarking on this journey of self-mastery and optimization.

Center. Expand. Flow.

I believe we have a responsibility as humans to seek evolution in order to live our lives at the highest possible expression of our beings. We have a responsibility to ourselves and to our fellow humans to seek this growth and evolution.

Chances are, if you're reading this, you're a truth-seeker of some sort. You're somewhere on your own journey to enlightenment. For that, I commend you. But I also challenge you to audit that self-optimization.

Ask yourself, are you growing and evolving as a complete human?

As humans, our goal should be to expand and evolve in a holistic way. To be a complete human, we must view our evolution as a 360-degree sphere, rather than growing unequally in certain domains or even expanding as a flat circle expanding in two dimensions.

It's not enough develop only our physical bodies. Likewise, an emotionally evolved human with a fragile body is not the highest expression of what it means to be human. To embody and express our existence at the highest level, me must cultivate health and evolve our minds, bodies, and spirits as if we are a sphere expanding in 360 degrees.

Even with this visual of a 360-degree sphere, we must seek to grow evenly and in all capacities, lest we become a lumpy orb rather than a well-rounded sphere that rolls true. Think about how a perfectly round ball might move through the world compared to an out-of-round orb – and it's no coincidence that "well-rounded" is a term used for educated, cultured, evolved humans. Continue meditating on this thought, and you'll see how becoming a well-rounded, more fully evolved being will allow us to move through the world with grace and fewer bumps, hiccups, and self-inflicted course detours.

The first step in realizing the performance we seek is realizing that performance is the manifestation, the outward expression, of underlying traits and capabilities.

In the physical world, the expression of power, balance, and agility one might see at the NFL combine or a Cirque Du Soleil show is merely the outward expression of the underlying health and abilities (skills) developed and honed through hours, weeks, and years of focused training.

Similarly, the focused, calm, Zen-like state demonstrated by a transcendental meditation master in the midst of rush-hour traffic in the busiest city is a quality that can only be expressed if it is first possessed – and possessed not by accident, but through dedicated skill acquisition practice.

Performance is the ultimate expression of our qualities, and these traits are built upon a foundation or center. We must first build this solid core before we can display its qualities.

Think about times you've moved through the world with ease. Now think about times you've struggled and notice the lack of center or balance in those times of struggle. Were you too far inward leaning? Were we looking too much to the external or outside world, failing to remain grounded in ourselves?

The impact of not being balanced, or centered, is a disruption in our ability to express ourselves at our fullest potential, and this results in energy blocks and a compromised ability to glide through the world.

Left unchecked, this can impact our cells, disease, gut health, blood, thoughts, feelings, moods, performance, and decisions.

Here is the formula for performance: Center. Expand. Flow.

1. Center – Who are you? What do you want? What is your authentic life?
2. Expand – How do we get there? What's missing? What is the gap and how do we close it?
3. Flow – Share this thing with the world. What's your medium? What's your message? Channel it and let it flow.

Centering

The center of the sphere is who we are and what we want. It is our authentic self.

By definition, anything whole must have a center. In the words of Lao Tzu, the man who penned *The Tao*

Te Ching ("The Way of the Way") hundreds of years ago, "If there is wholeness, there is center."

To be centered means to find balance between inward and outward.

When we look too much inward, we lose the ability innovate, learn, and grow. We can get tangled in our ego and allow self-doubt, fear, and lower levels of consciousness to creep in.

When we look too much outward, we forget that everything we need to achieve our dreams and live our authentic life is already inside of us. We get distracted by "shiny objects" and the lives of others.

Our work – and this is a lifelong pursuit – should be to strike this balance between inward and outward. One way to stay anchored in our center is to identify the core values that represent who we are.

These values will be different for each of us, but whatever they are – honesty, integrity, authenticity, strength, wisdom – in order to remain centered, it is essential that we identify and stay aligned with these values.

If you haven't identified these, I encourage you to think about four or five values that you want to live by. Let them be your guiding compass.

Understanding and staying aligned with our core values makes it easy for us to expand and flow in the right ways. It's not about reaching for or seeking results (outcomes); it is simply a matter of showing up, continuing to develop, learn and master our crafts, doing what is right, and being consistent with our *actions*.

If these actions are in line with our stated goals

and the goals built around our values, then we remain balanced and centered and keep developing our passion, and living life at the highest expression of our selves becomes second nature.

Let's be honest: those pesky negative emotions, feelings, and thoughts will not disappear completely. But they will reduce in both frequency and magnitude. They will be easier to identify and dismiss when they do appear.

Along with our core values, we'll likely find our passion here in our center – be it helping people, educating others, health, or art, your passion is the thing that sparks your inner self, and we're operating at our highest level of self when this passion is expanding and flowing from our center.

When we're aligned, there is almost no stopping this flow. This is why so many people say that it's not work when you do what you love.

On the other side of the spectrum are immobilization, fear, doubt, and procrastination. I will argue, however, that there is no procrastination – only misalignment, or being out of center.

Most procrastination is actually the avoidance of things that are misaligned with our passions and true selves. Even our procrastinations around our passions are derived from fear, self-doubt, or some other negative emotion that is: a) misaligned with the values we identified above and b) activating our defense response, lowering vagal tone, and spawning self-sabotaging feelings, thoughts, and behaviors.

Now that we're rooted, ready to be nourished and ready to flourish, let's talk expansion.

Expanding

A man who is the same at fifty as he was at twenty has wasted thirty years. – Muhammed Ali

The expanding component of the Center-Expand-Flow formula is predicated on our ability to remain rooted in our values. The tighter that connection, the more genuine and authentic our thoughts and actions become.

As we said above, what is rooted can easily be nourished and then flourish.

Expansion is all about nourishing – moving from our current level to the next levels. It's about developing our passions, mastering ourselves and our crafts, and creating the legacy we want to leave behind. It's about becoming and living as the highest expression of our self.

This can mean our level of vibration in an acute setting, elevating our knowledge or our emotional intelligence. Remember: expansion should occur in a 360-degree sphere, not just one area or dimension.

The key to expansion is to maintain an objective, non-judgmental look at where we are and identify the gaps between the place we currently occupy and the place wish to occupy.

Questions we must ask and answer as we seek to expand: how do we get there? What's missing? What is the gap and how do we close it?

Expansion is predicated on seeking knowledge – having a growth mindset, seeking that which we lack – not in material possessions, but in knowledge – so that we can close the gap between where we are now and where we want to be.

Expansion can take place in an acute setting, by consciously choosing to elevate our level of vibration during a rough morning, and it should also be part of our long-term growth goals – to grow and evolve as humans month over month, year after year.

Recall our 360-degree sphere and realize that this expansion should occur in all disciplines of our life, including physical performance, mental performance, relationships, emotional and spiritual well-being, knowledge, awareness, and consciousness.

Keep in mind that expansion is limited by poor vagal tone. When we feel threatened, we shrink, seeking security, and we play small.

Flowing

Think of Center-Expand-Flow as a pyramid, with each level being built on the foundations below it. Center is the bottom level, Expand is the middle level, and Flow is the tip of the pyramid.

If centering is the rooting, and expansion is the nourishing, then flow is the flourishing.

Like our discussions of performance, flow is the outward expression of the abilities, skills, and operating systems we have in place. As such, it depends on and can be limited by levels.

Flowing is our interface with the world. It's the art we create, the businesses we build, and the relationships we cultivate.

What's your mission? What's your message? Share this thing with the world. What's your medium? Channel it and let it flow.

The more our life is built around our passions, the more we develop those passions and create mastery, the more flow we can enjoy in our life. The more we solidify the foundation layers of this pyramid, the greater and more enjoyable our expression/performance can become.

When I say flow, I'm not necessarily referring to the flow states that have become in vogue recently. Instead, I'm referring to our ability to channel what is inside of us and bring it forth into the world, to be and live as the person we know we want to be.

Sure, one reward of this aligned expression is to experience more of the flow-like peak states, but I caution that these peak states, by definition, are not sustainable. They also have downsides in their aftermath.

Rather than chase peak performance, I posit that we seek high performance on a consistent basis. By definition, a peak has a short-lived, transient existence, and is followed by a drop-off.

Instead of spending our lives chasing one high point to another high point, we should seek to elevate our baseline so that our "normal" is actually a higher standard than what is commonly seen in the world.

Normal and average have become bastardized. If we build our lives around our passions, constantly expand our abilities and consciousness, and let our choices flow from this center, we'll experience this sustainable high level of performance that elevates us beyond the status quo of mediocrity and settling.

Don't get me wrong, peak performance still has its place and should be called upon for optimal living.

But peaking should be used sparingly, and only for special occasions a few times a year or month – not relied upon daily.

Stop waiting for permission to be you; do it and do what makes you happy. Don't focus on flow. Flow happens effortlessly when we nail the first two levels of this formula. Stay centered in your values, develop and master your passion, and build your daily habits around this. Flow and high-performance living are the *outcome* that you'll experience by focusing on the *actions* of the first two levels.

While we're talking about not waiting for permission, realize that in today's world, more than ever, we initiate and iterate first, then validate later. Look at all the YouTube celebrities making millions of dollars online who never asked for permission to start. They created content, helped or entertained hundreds, then thousands, and ultimately millions by developing their passion.

What we see once they have accumulated enough "success" for the world to realize they have one million subscribers is the resulting flow of their passion and creativity that stems from their action-oriented focus on centering and expanding.

Was their following that large on day one? No. Was the video quality and camera presence perfect in their first few videos? No.

But they did the best they could for their current level of development, and then continued to master their message and their craft. They were consistent, aligned, and constantly growing (expanding).

How can we embody this Center-Expand-Flow formula?

Take the values you wrote down in the Center section and keep them handy. Take the three words that describe the highest version of yourself from earlier. Set that reminder on your phone to act as if you were that person.

Move through the world as this person. Use your values to weigh and judge every opportunity, keeping yourself centered and in alignment with your vision.

Block off time on your schedule each day to work on your passion – write, draw, film, photograph, speak, or volunteer.

Make continuing education a value in some form. Time plus repetitions leads to mastery of passion.

Most people need more mastery, more focused, quality, positive repetitions, not more information. One key from Kotler and Wheal in *The Rise of Superman* is that focus is a prerequisite for flow. This holds true for high performance, peak states, and elevated awareness.

Grow. Evolve. Honor the responsibility that we have to be intentional about our life, our health, our self-evolution, and the way in which we move through this world.

Be the master of your ship, captain of your soul.

Be in control. Don't let your subconscious run the show.

F*ck Your Feelings. Move the Chains.

The fear, the negative feelings, and the darkness in the shadows will always be there.

The voices in our head, the self-doubt, the demons

– whatever your personal demons may be – will always be there.

Make no mistake, I'm not making the promise that you'll banish them for eternity. Rather, think of these unwanted emotions/demons as existing on a spectrum.

Our goal is to become aware of them – aware of how they impact our lives and devise strategies to mitigate their negative influence.

We can't avoid feelings, and we wouldn't want to – lest we become like Damasio's patient Eliot, incapable of making even the simplest decision – so we seek to understand and enhance our primitive operating system, strengthening our neural pathways to higher levels of cognition and consciousness, and create better defaults for ourselves so that we can move from our current level to higher levels of existence.

We can't ignore feelings; as Darwin noted, they're an evolutionary adaptation.

But we *can* get better at being aware of how they may be inadvertently directing our lives. Using the methods presented in this book, you're now equipped to elevate your consciousness, increase your awareness, create choice – better default choices – for yourself and ready to define your values and make sure your actions are congruent with those values and your goals.

It's one thing to read this book. The rub is in the implementation. Knowledge without implementation is the same as ignorance – maybe worse.

1. What are your three to five big takeaways from this book? (I'd love to hear them! Post them

on social and tag me.)
2. How will you implement those takeaways?

Activities

I'd like to close this book by saying thank you. Our time is our most precious resource, and I'm honored you gave me and my thoughts this much of your time and focus.

I truly hope that if you're reading at this point, this book has been of great value to you.

I'm fascinated by why some people succeed and live their dream lives, while others do not. Throughout my professional career – be it personal trainer, nutritionist, gym owner, writer, or podcaster – my mission has always been to help everyone I encounter lead a stronger, happier, healthier life for having crossed paths with me.

This book is the latest project in that effort.

As we've seen, successful people are those who are able to master the space between their ears, act in alignment with their values, and prevent their feelings from running the show and leading them astray.

You now have the knowledge you need to understand and control your biology.

You have the tools you'll need to implement your plan and create the future you want for yourself.

What you do with them is up to you. Knowledge without implementation is wasted knowledge.

Like my SEALFIT instructor, I'm now the one drawing the line in the sand.

Will you realize your potential for greatness? Who are you? Where do you stand? What will you do with this life you've been blessed with? What impact will you make? What will your legacy be?

Define your values. Act as if you already are that person. Become that person. Master your mind and control your own future.

F*ck those feelings that try to derail you along the way.

One Final Thought

A great coach once told me, "If you want more, you need to help more people." I've tried to keep that lesson at the core of all that I do, and it's been remarkably accurate advice.

It's also made everything I do more rewarding. Helping others achieve what they want in life is the greatest feeling I know. Perhaps you've experienced this as well.

I have a request of you. If you've enjoyed this book and found value in it, consider giving a copy to three people for whom you care about. They can be friends, family, co-workers, business partners, or anyone you want to help.

Who do you know that will benefit from the message in this book?

1. _____
2. _____
3. _____

I know this sounds like it benefits me and my

mission to reach and help as many people as possible, so we can make a greater impact on this world. It does. But it also benefits you and your community. You're helping the people around you – those you care about most – elevating and upgrading their lives. That is the powerful thing we can do as humans. This book could forever alter the course of someone's life, and it could be you who gives it to them.

Thank you for the valuable time you devoted to reading this book. I'm grateful to be a part of your journey and look forward to helping you in any way I can. Please email me at ryan@ryanmunsey.com if there is anything I can do to help you.

I love hearing your success stories. Connect with me on social media and share your story.

I'm most active on Instagram @ryanmunsey_

If you post a photo of yourself and the book, we'll share it and your success story. Make sure to tag me (@ryanmunsey_) in the photo.

APPENDIX 1: My Nine Go-To "Tools" to F*ck My Feelings

I'll close this book with an appendix of sorts. Diet books have recipes in the back, fitness books include workouts, so I'd like to include my personal selection of "go-to tools" to change my physiology on those days I just don't have that spark.

Let's be honest: we all have them. They're part of this amazing journey we call life.

But sometimes we have put our big boy (or girl) britches on and get on with our lives even when we don't feel like it.

It's frustrating, I know. We do everything right, yet those days sneak up and bite us when we're least expecting it – or worse, on a day when we can least afford to be off our game.

Here's how I manage to have solid days – sometimes immensely productive – days even when I don't feel like it.

ONE: #MTC. Move the Motherf*cking Chains

I'm a big list person. I have lists on my phone, computer, in my office, and in the kitchen – my mind is always churning on something, and I have to write things down if they're ever going to have a chance to survive.

One of those lists is my yellow legal pad that serves as my "to-do list" for each day.

The top right quadrant of each page has #MTC. Under that, I write the single most important activity for that day. This is the one thing that, no matter what else happens, needs to get accomplished on that day.

If at all possible, I attack that item first thing in the morning. Sometimes it's an appointment, podcast, or phone call that has to wait, in which case, the timing is out of my control.

I know that accomplishing that one item will move my mission forward, and I make sure that one thing gets done – and gets done with my full presence and intention.

TWO: Music

Nothing, absolutely nothing alters my mood as quickly or as powerfully as music. I'll never have the opportunity to experience this as another person, it's forever an n=1 experience for me, so I'm not sure if my response is human or individual.

But with the popularity of music and our rich tradition of music throughout history, I'm betting music will be a mood-hack for you too.

The secret it to change it up. Some days Rage Against the Machine gets me where I need to be. Other days it is Van Morrison radio on Pandora, Laid Back Beach Music, or even Mozart. Sometimes, it's Motown, Lyric-Free EDM or Trance, or lately my Prince/ MJ/80's station on Pandora.

No two days are the same, and sometimes it's a process to find the right one. Change it up, try different options, but one caution: don't spend all your focus picking the right music. If you try three one day and it's not happening, move on to a different "hack."

Don't lose sight of the fact that the music is there to enhance your work – you're getting paid to DJ your own day.

THREE: "I'm f*cking awesome"

It's nearly impossible for me to keep a straight face when I look into the mirror, hold a power pose and say "I'm f*cking awesome."

But I say it. And then I say it again. And again, three to five times, or until I start to laugh and/or believe it. Either way, my state is different than it was thirty seconds before.

Sub in whatever affirmations you want, but give this one a try too.

These work, and they work quickly.

FOUR: Play

We talked about this one in the book, so I won't go into great detail – just break up your routine and get out of your head. Go somewhere new, doing something spontaneous. Make it unstructured (don't time it, track it with a Fitbit, or measure it) and act like a kid.

Surf, run, roll down hills, play in the dirt, go to the gun range, play pickup games at the park (basketball, soccer, Frisbee – just make sure you're present and that you lose yourself in the activity.

You'll be amazed at how you feel on the other

side.

FIVE: Text Five People

This is another piece of sagely advise from my mentor Paul Reddick. I was being a bitch and complaining about my feelings, so he told me to stop focusing on myself, focus on helping others, and simply text five people every morning for a week.

The rules are simple:

1. Text five people in the morning.
2. Don't ask them how they're doing or some other non-sense. Send them a funny image, a sincere note, or something else unique that will make their day.

It's profound. Not only will it enhance your mood and your productivity for that day, but you'll make the world a better place and positively impact five other people. Imagine what earth would be like if everyone started their day like this.

SIX: Take a Cold Shower

Yes, this is usually part of my morning routine. Sometimes, if I'm struggling and nothing else seems to work, I do it again. The simple act of choosing to do something uncomfortable builds momentum and makes subsequent tasks easier to start. Starting is always the toughest part.

We know our feelings dictate our decisions. Exercising our ability to do the hard thing, despite our feelings about wanting to stay comfortable, cultivates our neural pathways involved in decision making. It's like bicep curls for your "decision-making muscles."

SEVEN: Gratitude

I'm not talking about the usual gratitude journaling here. If we're really struggling one day, then journaling isn't going to help. We need to really feel it. Nothing shocks me out of a state of self-pity like visiting or volunteering for the homeless or for veterans. Sometimes we need a serious jolt to remind ourselves how good we have it and that we're wasting an opportunity that others would kill for.

EIGHT: Float

If these feelings persist, or my brain feels overwhelmed – on information overload – I know it's time to float and process all that data.

Other methods of consciousness exploration could also be employed to help us see the bigger picture. You know like breathing, yoga, meditation, et al.

NINE: Remind Myself of my Personal Agreement (The Darkness Contract)

I call it "The Darkness Contract" because it's a contract I made with myself after watching the making of *Darkness on The Edge of Town*, Bruce Springsteen's 1979 album (and my favorite album of all time).

Here is it. Some of these words are directly from Bruce. I never intended this to be something I published, but I've shared it with some coaching clients and they loved it. Maybe it'll help you too. Feel free to use this or create your own.

The Darkness Contract

Tonight I'll be on that hill 'cause I can't stop
I'll be on that hill with everything I got
With our lives on the line where dreams are found and lost
I'll be there on time and I'll pay the cost
For wanting things that can only be found
In the darkness on the edge of town.
– lyrics from *Darkness on the Edge of Town*

It's a personal reckoning.

It's about resilience, and it's a commitment *to* life – *your* life.

How do I honor the breath in my lungs and this life I've been given?

With whom will you stand?

For what will you stand?

What will you tolerate of yourself, for yourself, for this one life you get to pursue and enjoy?

Take ownership, accountability, and control – it's your ass, your legacy on the line.

Live as if you were writing your own biography or your own eulogy.

Who will brave the rain to watch your burial? Will you be missed? Did you make an impact? What will be your legacy?

There is something inside you – you're one of the fortunate, lucky few who has identified that authentic self and true life's calling. Many people go their whole lives without identifying that thing. Man, you're lucky.

Do not forsake your own inner life force.

Now sack the f*ck up, quit feeling sorry for yourself, and make shit happen because nobody else is going to do it for you.

And never quit. It won't be easy. It won't happen the first try – you're going to have to work and fight for it every day of your life – but what better way to spend your life than in pursuit of the thing that brings you the most joy, satisfaction, and fulfillment? Spend your life pursuing it. Bring it to life. You can't lose that thing (see the three people in your head).

If it ever gets to be too much to handle, call Dad, and we'll visit a Children's Cancer/AIDS hospital together. *

* My Dad has a friend whose son committed suicide. When he told me the story, he asked me to do this for him, if I ever felt as though suicide was the only way out. No matter how bad it gets, someone would love to have your problems. There are always people happier with less or in worse conditions.

APPENDIX 2: Putting it All Together – The Perfect Day

This is an example of a possible daily routine that includes as many of our trait-developing habits and state-altering tools as possible.

A note on routines. Eventually, by definition, they become mundane routines. This should never be a checklist of things you have to do. If you find yourself going through the motions or performing these without intention, stop and change it up.

If we ever get to a place where we feel like we cannot be our best self without performing a routine, we need to re-evaluate our relationship with said routine. That mindset gives the power to the routine, not us.

Also, we need to understand the difference between doing what we don't feel like because we've committed to it and doing something unnecessary just because we think we have to. The audit for this is to ask yourself: if this working? Is this getting me closer to my goal? If yes, keep it. If no, ditch it and look for something better.

1. 5,4,3,2,1 – get up. Never hit snooze again.

Keep your alarm out of reach from your bed. This forces you to physically get out of bed to turn it off. If you need more sleep, go to bed earlier. The habit of not hitting snooze has changed my life.

2. Get outside. Do some grounding and get some morning sun within one hour of sunrise (no glasses or lenses of any kind).
3. You can blend the above step with meditation/breathing or do here as stand-alone.
4. Take a cold shower. (Steps #1-4 can be completed within twenty to thirty minutes or less of waking up every single day.)
5. Accomplish your *one thing* for the day. Move the Chains.
6. Move. There is no bad time to get movement into to your day, but this is as good a place in your lineup as any. You'll build on the momentum you've already created and set yourself up for a great remainder of the day.
7. Practice gratitude. Like movement, gratitude's benefits are independent of the time of day you do it. If this is a new habit, place it at a time of day where you can be consistent with it. The evenings right before bed is a great time, plus research has shown it helps sleep quality.
8. Develop a consistent sleep routine. Avoid blue light after dark, no screen or technology thirty minutes before going to bed, read, do your gratitude journal, or do breathe work. Whatever you choose, be consistent. (For that

reason, less is more.)
9. Throughout the day, stay aware of posture and light from screens. Use screen-filtering programs like iris or f.lux, avoid staring at the same screen for longer than twenty to thirty minutes at a time. Avoid sitting in the same position for long periods of time as well. Be aware of how your environment and position impact your physiology – and, in turn, your feelings.

"Aces" up your Sleeve

Not every day is a 10/10. Some days we need a little extra boost. Pull these tools out when you need a quick, transient boost for your state.

<u>Movement:</u> Set a timer for one minute and choose an exercise – pushups, burpees, plank, pull-ups. Don't stop until the minute is over. Or go outside and walk for fifteen or twenty minutes. This is my go-to to clear my head, focus my thoughts and recharge.

<u>Music:</u> Blast some Springsteen, dance to 80s radio, or anything in between. Sometimes this is all it takes. Combine music and movement by playing the song Thunderstruck from AC/DC. Every time they say the word "thunder," do a burpee.

<u>Breathing:</u> Go back to Chapter 7 and choose the box breathing or Wim Hof method and spend a few minutes focusing on the oxygen that gives us life.

<u>Play:</u> I still think back to that day I found a lonely basketball and an eight-foot rim and how much joy it brought me to be carefree for ten minutes. Roll around on the floor with an animal, play with your kids (or cousins, nieces, nephews, etc.) – find something that is pure fun and immerse yourself in it. The key is to lose yourself in the activity.

<u>Meditation:</u> Close your eyes and just be. Let go of time, desires, and all the thoughts in your head. Just breathe and be still for five minutes. You have the time. If you think you don't, you need this more than you realize.

<u>Gratitude/attitude adjustment:</u> When we check ourselves by reminding our conscious brain that we have both arms, both legs, and our sight – when many other do not – it's tough to remain in a state of self-pity. This doesn't always work, but nine times out of ten, this self-check is enough to alter our state in the right direction. Count your blessings, especially the ones we usually take for granted. Like the fact we woke up today.

APPENDIX 3: A Few of My Favorite Books and Resources

Books

Growth Mindset, Carol Dweck
The Code of the Extraordinary Mind, Vishen Lakhiani
The Compound Effect, Darren Hardy
Tao Te Ching (I prefer the Stephen Mitchel translation)
Originals, Adam Grant
Start with Why and *Leaders Eat Last*, Simon Sinek
Extreme Ownership, Jocko Willink and Leif Babin
Go Right, Logan Gelbrich
The Blue Zones, Dan Buetner
Spark and *Go Wild*, John J. Ratey
Supple Leopard and *Ready to Run*, Kelly Starrett
Original Strength, Tim Anderson
War of Art, Turning Pro, and *Do the Work,* Steven Pressfield
Bold and *Abundance*, Peter Diamandis
Epi-Paleo Rx, Jack Kruse
Deep Nutrition, Dr. Cate Shanahan
The 150 Healthiest Foods on Earth, Jonny Bowden

Amateurs do it till they get it right. Pros do it until they can't get it wrong. — Steven Pressfield

Other Resources

Holosync, Bill Harris
Peak Brain Institute and Dr. Andrew Hill
Found My Fitness, Dr. Rhonda Patrick
Evolution Human Performance, Scott Dolly
Rick Alexander
Aaron Alexander (no relation)
Wim Hof

Acknowledgements

I am an ardent believer in the saying that no one does it alone. My life and this book are no exception.

I offer my most sincere appreciation to:

My wife Donna has been my biggest supporter from day one. In 2012, she was the one who gave me the final nudge to open my own gym, thus launching my entrepreneurial career (a nudge she may regret as any entrepreneur and their family can understand).

To Donna: thanks for your unwavering support, patience, and understanding. I love you.

My parents, coaches, mentors, and Paul Reddick for helping me become the person I am today. My efforts to help others and improve the world are a direct reflection of the lessons you've taught me and the impact you've made in my life.

The high performers, researchers, scientists, and experts who leant experience and expertise to this book. Your willingness to share knowledge for the benefit of others is greatly appreciated.

My friends, book team, editor Jonathan Green, designer, and my entire professional community for the support, sharing of knowledge, and inspiration to create and release this book.

Ryland Hormel, my co-founder at The Better Human Project and media director for this book has been a huge support and teacher.

Thank you all. And thank YOU for reading.

References

1. www.ted.com/speakers/antonio_damasio.
2. Henley W.E., *Invictus*.
3. Damasio A.R., *Descarte's Error: Emotion, Reason and the Human Brain*, 2005.
4. www.boundless.com/psychology/textbooks/boundless-psychology-textbook/biological-foundations-of-psychology-3/structure-and-function-of-the-brain-35/the-limbic-system-154-12689/.
5. www.sciencedaily.com/terms/limbic_system.htm.
6. www.technologyreview.com/s/528151/the-importance-of-feelings/.
7. www.psychologytoday.com/blog/one-among-many/201006/reason-and-emotion-note-plato-darwin-and-damasio.
8. Anderson A.K., Affective Influences on the Attentional Dynamics Supporting Awareness, *Journal of Experimental Psychology-General*, 2005, vol. 134 (pg. 258-281).
9. Anderson A.K. (email interview), 2017.
10. Cannon W., The James-Lange Theory of Emotions: A Critical Examination and an

Alternative Theory, *The American Journal of Psychology*, 1927, 39: 106-124.
11. www.en.wikipedia.org/wiki/William_James.
12. Schioldann J., On Periodical Depressions and Their Pathogenesis by Carl Lange (1886), *History of Psychiatry*, 2011, 22 (85 Pt 1): 108–130.
13. www.en.wikipedia.org/wiki/James%E2%80%93Lange_theory.
14. Barrett L.F., and Niedenthal P.M., Valence Focus and the Perception of Facial Affect, *Emotion*, 2004, vol. 43 (pg. 266-74).
15. Anderson A.K., Affective Influences on the Attentional Dynamics Supporting Awareness, *Journal of Experimental Psychology-General*, 2005, vol. 134 (pg. 258-281).
16. Rowe G., Hirsh J.B. and Anderson A.K., Positive Affect Increases the Breadth of Attentional Selection, *Proceedings of the National Academy of Sciences of the United State of America*, 2007, vol. 104 1 (pg. 383-8).
17. Adam K. and Anderson A.K.; Feeling Emotional: The Amygdala Links Emotional Perception and Experience, *Social Cognitive and Affective Neuroscience*, Volume 2, Issue 2, 2007 (pg. 71-72).
18. MacLean P.D., The Triune Brain in Conflict, *Psychotherapy Psychosomatics*, 1977; 28:207-220.
19. Darwin C., *On the Origin of Species by Means of Natural Selection, or the Preservation of Favoured Races in the Struggle for Life*, 1859.
20. Dawkins R., *The Selfish Gene*, 1976.

21. www.marctomarket.com/2013/08/great-graphic-your-brain-before-and.html.
22. www.creativitypost.com/education/the_benefits_of_movement_in_schools.
23. www.ncbi.nlm.nih.gov/pmc/articles/PMC3674785/.
24. www.psychologytoday.com/blog/the-human-beast/200910/how-much-physical-activity-do-we-really-need.
25. Cordoin L. et al., Physical Activity, Energy Expenditure and Fitness: An Evolutionary Perspective, *International Journal of Sports Medicine*, 1998, 10 (pg. 328-335).
26. Levine, J.A., Non-Exercise Activity Thermogenesis, *Proceedings of the Nutrition Society*, 2003, 62 (pg. 667-679).
27. www.mayoclinicproceedings.org/article/S0025-6196(16)00043-4/fulltext.
28. www.ajmc.com/newsroom/few-americans-follow-all-4-elements-of-a-heart-healthy-lifestyle.
29. Buettner D., *The Blue Zones*, 2012.
30. www.apa.org/pubs/journals/releases/xlm-a0036577.pdf.
31. www.ncbi.nlm.nih.gov/pmc/articles/PMC3674785/.
32. Blumenthal J.A. et al., Effects of exercise training on older patients with major depression, *Arch Intern Med*, 1999 Oct 25; 159(19):2349-56.
33. Ratey J., *Spark: The Revolutionary New Science of Exercise and the Brain*, 2013.

34. www.ncbi.nlm.nih.gov/pubmed/21934728.
35. www.sciencedirect.com/science/article/pii/S1074742706001596.
36. www.sciencedirect.com/science/article/pii/S2095254614001161.
37. www.ncbi.nlm.nih.gov/pmc/articles/PMC4563312/.
38. www.chriskresser.com/9-steps-to-perfect-health-7-move-like-your-ancestors/.
39. McCraty R, Shaffer F. Heart Rate Variability: New Perspectives on Physiological Mechanisms, Assessment of Self-regulatory Capacity, and Health risk. *Global Advances in Health and Medicine*. 2015;4(1):46-61. doi:10.7453/gahmj.2014.073.
40. Matthews C.E. et al., Amount of Time Spent in Sedentary Behaviors in the United States, *American Journal of Epidemiology*, 2008;167(7):875-881.
41. www.sittingissmoking.com/sitting_is_the_new_smoking.pdf.
42. www.washingtonpost.com/apps/g/page/national/the-health-hazards-of-sitting/750/.
43. www.sittingissmoking.com/sitting_is_the_new_smoking.pdf.
44. Cuddy A.J.C., Wilmuth C.A. and Carney D.R., *The Benefit of Power Posing Before a High-Stakes Social Evaluation*, Harvard Business School Working Paper No. 13-027, 2012.
45. www.ted.com/talks/amy_cuddy_your_body_language_shapes_who_you_are.
46. www.cdc.gov/physicalactivity/basics/pa-

health/index.htm.
47. Marquez J.L. et al., Cyclic Hypobaric Hypoxia Improves Markers of Glucose Metabolism in Middle-Aged Men, *High Altitude Medicine and Biology*, 2013;14(3):263-72.
48. Mischel W. Et al., Cognitive and Attentional Mechanisms in Delay of Gratification, *Journal of Personality and Social Psychology*, 1972, 21 (2): 204-218.
49. Rodrick Wallace, Columbia University professor and author of *Computational Psychiatry: A Systems Biology Approach to the Epigenetics of Mental Disorders*, reminds us, "Emotions are self-regulatory responses that allow efficient coordination of the organism for goal-oriented behaviors."
50. Błachnio A. and Przepiorka A., Dysfunction of Self-Regulation and Self-Control in Facebook Addiction, *The Psychiatric Quarterly*, 2016;87:493-500.
51. www.link.springer.com/article/10.1007/s11126-015-9403-1.
52. Kotler W., *Stealing Fire*, 2017.
53. Kotler W., *The Rise of Superman*, 2014.
54. Barrett L.F., Being Emotional During Decision Making – Good or Bad? An Empirical Investigation, *Academy of Management Journal*, 2007;50(4):923-940.
55. Eagleman D, *Incognito: The Secret Lives of the Brain*, 2012.
56. Lakhiani V., *Code of the Extraordinary Mind*, 2016.

57. Bill Harris (interview for Optimal Performance podcast), 2016.
58. Kolb B. et al., Experience and the Developing Prefrontal Cortex, *Proceedings of the National Academy of Sciences of the United States of America*, 2012;109(Suppl 2):17186-17193.
59. www.link.springer.com/article/10.1007%2Fs10602-008-9056-2.
60. www.ncbi.nlm.nih.gov/pubmed/5010404.
61. www.jneurosci.org/content/35/38/13194.
62. www.academic.oup.com/sleep/article/36/12/1919/2709417/Circadian-Variation-of-Heart-Rate-Variability.
63. www.ncbi.nlm.nih.gov/pubmed/18437004.
64. www.ncbi.nlm.nih.gov/pmc/articles/PMC3797399/.
65. www.youtube.com/watch?v=xiXZVDKRe00.
66. www.journals.plos.org/plosone/article?id=10.1371/journal.pone.0033769.
67. www.sciencedirect.com/science/article/pii/S0022399997000044.
68. www.ncbi.nlm.nih.gov/pmc/articles/PMC3747835/.
69. www.ncbi.nlm.nih.gov/pmc/articles/PMC2769007/.
70. www.healthysleep.med.harvard.edu/need-sleep/whats-in-it-for-you/mood.
71. www.healthysleep.med.harvard.edu/healthy/matters/benefits-of-sleep.
72. www.academic.oup.com/aje/article/160/6/521/79596/Sleep-Disordered-

Breathing-Glucose-Intolerance-and.
73. www.healthysleep.med.harvard.edu/healthy/matters/benefits-of-sleep.
74. Dinges D.F. et al., Cumulative Sleepiness, Mood Disturbance and Psychomotor Vigilance Performance Decrements During a Week of Sleep Restricted to 4-5 Hours Per Night, *Journal of Sleep Research and Sleep Medicine*, 1997, 20, 267-277.
75. Simon, H. Sleep Helps Learning, Memory. Harvard Health. February 2012.
76. www.pnas.org/content/113/26/7272.full.
77. Wood A.M. et al., Gratitude Influences Sleep Through the Mechanism of Pre-Sleep Cognitions, *The Journal of Psychosomatic Research*, 2009 Jan; 66(1):43-8.
78. www.sleep.org/articles/temperature-for-sleep/.
79. Darwin C., *The Expression of The Emotions in Man and Animals*, 1872.
80. www.blog.bulletproof.com/wp-content/uploads/2015/11/transcript-stephen-porges-264.pdf.
81. Berthoud H.R. et al., Functional and Chemical Anatomy of the Afferent Vagal System, *Autonomic Neuroscience: Basic and Clinical*, Volume 85, Issue 1 (pg. 1-17).
82. My interview with Porges: www.ryanmunsey.com/142-dr-stephen-porges-on-hrv-and-polyvagal-theory/.
83. Hisako T. et al., Impact of Reduced Heart Rate Variability on Risk for Cardiac Events,

Circulation, 1996, 94:2850-2855.
84. www.ncbi.nlm.nih.gov/pubmed/18003665.
85. www.tandfonline.com/doi/abs/10.1080/08870440290025867.
86. www.ncbi.nlm.nih.gov/pubmed/19424767.
87. www.ncbi.nlm.nih.gov/pubmed/22178086.
88. www.ncbi.nlm.nih.gov/pubmed/8425692.
89. My interview with Rick Strassman: www.naturalstacks.com/blogs/news/rick-strassman-pineal-gland-dmt-and-consciousness.
90. www.hindawi.com/journals/ecam/2014/819871/.
91. www.businessinsider.com/neuroscientist-most-important-choice-in-life-2017-7.
92. www.ncbi.nlm.nih.gov/pmc/articles/PMC3510904/.
93. www.newsroom.ucla.edu/stories/marco-iacoboni-mirror-neurons.
94. www.nytimes.com/2016/02/28/magazine/what-google-learned-from-its-quest-to-build-the-perfect-team.html.
95. Park G. and Thayer J.F., From the Heart to The Mind: Cardiac Vagal Tone Modulates Top-Down and Bottom-Up Visual Perception and Attention to Emotional Stimuli, *Frontiers in Psychology*, 2014, 5:278.
96. Charlie Hoehn on OPP: www.ryanmunsey.com/113-play-away-your-anxiety-with-charlie-hoehn/.
97. Maslow A.H., *Religions, Values and Peak Experiences*, 1964.

98. www.yougenics.net/griffis/courses/arts344/readings/exploratoryPlay.pdf.
99. www.hbr.org/2015/12/calming-your-brain-during-conflict.
100. www.whatisessential.org/sites/default/files/Neurobiology%20of%20Threat.pdf.
101. www.journals.sagepub.com/doi/abs/10.1177/106342660000800405.
102. www.timeshighereducation.com/features/the-importance-of-play/2012937.article#survey-answer.
103. www.apa.org/pubs/journals/releases/bul-1316803.pdf.
104. www.scientificamerican.com/article/smile-it-could-make-you-happier/.
105. www.science.sciencemag.org/content/313/5787/684.
106. www.psychologytoday.com/blog/what-mentally-strong-people-dont-do/201504/7-scientifically-proven-benefits-gratitude.
107. www.link.springer.com/article/10.1007/s10943-015-0063-0.
108. www.psycnet.apa.org/record/2003-01140-012.
109. www.psychology.as.uky.edu/gratitude-antidote-aggression.
110. www.ournals.sagepub.com/doi/abs/10.1177/1948550611416675.
111. Wood A.M. et al., Gratitude Influences Sleep Through the Mechanism of Pre-Sleep Cognitions, *Journal of Psychosomatic Research*, 2009, 66(1):43-8.

112. www.ncbi.nlm.nih.gov/pmc/articles/PMC3797399/.
113. www.ncbi.nlm.nih.gov/pubmed/18437004.
114. Grant A., *Originals: How Non-Conformists Move the World*, 2016.
115. www.huffingtonpost.com/2013/08/23/volunteering-happiness-depression-live-longer_n_3804274.html.
116. www.en.wikipedia.org/wiki/Enteric_nervous_system.
117. www.ncbi.nlm.nih.gov/books/NBK6273/.
118. www.fr.wikipedia.org/wiki/Pascal_Picq.
119. www.simplypsychology.org/maslow.html.
120. www.scientificamerican.com/article/gut-second-brain/.
121. Mayer E.A., Gut Feelings: The Emerging Biology of Gut-Brain Communication, *Nature Reviews Neuroscience*, 2011, 12(8):10.
122. www.verywell.com/what-is-a-neurotransmitter-2795394.
123. Sender R., Fuchs S. and Milo R., Revised Estimates for the Number of Human and Bacteria Cells in the Body, *PLoS Biology*, 2016, 14(8):e1002533.
124. Bonaz, B. Brain-Gut Interactions in Inflammatory Bowel Disease. American Gastroenterology Association. January 13, Vol. 144 DOI: http://dx.doi.org/10.1053/j.gastro.2012.10.003.
125. www.hrvcourse.com/case-study-the-effects-of-diet-on-heart-rate-variability/.
126. www.drsircus.com/cardiovascular/vagus-

nerve-inflammation-heart-rate-variability/.
127. www.myithlete.com/heart-rate-variability-health-pt-1-inflammation/.
128. www.onlinelibrary.wiley.com/doi/10.1111/j.1365-2796.2010.02321.x/full#js-feedback.
129. Parekh P.J. et al., The Role of Gut Microflora and the Cholinergic Anti-inflammatory Neuroendocrine System in Diabetes Mellitus, *Frontiers in Endocrinology*, 2016, 7:55.
130. www.health.harvard.edu/diseases-and-conditions/the-gut-brain-connection.
131. www.faculty.washington.edu/chudler/cells.html.
132. Beninger R., The Role of Dopamine in Locomotor Activity and Learning, *Brain Research Reviews*, Volume 6, Issue 2, October 1983 (pg. 173-196).
133. www.naturalstacks.com/blogs/news/117543237-neuroplasticity-train-your-brain-and-be-smarte.
134. Bressan R.A. and Crippa J.A., The Role of Dopamine in Reward and Pleasure Behaviour – Review of Data From Preclinical Research, Acta Psychiatrica Scandinavica, 2005, (427):14-21.
135. www.ncbi.nlm.nih.gov/pubmed/8548806.
136. Angier N.A., Molecule of Motivation, Dopamine Excels at Its Task, *NY Times*. October 26, 2009.
137. Asociación RUVID, Dopamine Regulates the Motivation to Act, Study Shows,

ScienceDaily, 10 January 2013.
138. www.today.uconn.edu/2012/11/uconn-researcher-dopamine-not-about-pleasure-anymore/.
139. Fox G.R. et al., Neural Correlates of Gratitude, *Frontiers in Psychology*, 2015, 6:1491.
140. www.ncbi.nlm.nih.gov/pmc/articles/PMC2290997/.
141. www.ncbi.nlm.nih.gov/pmc/articles/PMC3819153/.
142. www.muscle-health-fitness.com/natural-dopamine.html/.
143. Lehrer L., *How We Decide*, 2010.
144. Mattson M.P., Superior Pattern Processing is the Essence of the Evolved Human Brain, *Frontiers in Neuroscience*, 2014, 8:265.
145. Kurzweil R., *How to Create a Mind: The Secret of Human Thought Revealed*, 2012.
146. www.ncbi.nlm.nih.gov/pubmed/14998098.
147. www.quora.com/How-is-dopamine-involved-in-working-memory.
148. www.hrplab.org/brain-training-for-over-50s/.
149. www.hrplab.org/review-brain-training-depression-anxiety/.
150. www.hrplab.org/brief-review-of-2014-2015-studies-on-working-memory-training-for-iq-and-working-memory/.
151. www.authors.library.caltech.edu/56514/.
152. www.ncbi.nlm.nih.gov/pubmed/17241888.
153. www.ncbi.nlm.nih.gov/pmc/articles/PMC4764485/.

154. www.ncbi.nlm.nih.gov/pubmed/12480364.
155. www.versatables.com/discover/sit-and-stand-movement/sedentary-lifestyles/sedentary-lifestyles-linked-to-depression-anxiety/.
156. www.ncbi.nlm.nih.gov/pubmed/27491067.
157. My interview with Ferit: www.naturalstacks.com/blogs/news/how-to-target-food-intolerance-with-the-pinner-test.
158. http://www.sciencedirect.com/science/article/pii/S0016508504003798.
159. www.nature.com/nrgastro/journal/v10/n8/full/nrgastro.2013.105.html?foxtrotcallback=true.
160. Hunter P., The Inflammation Theory of Disease: The Growing Realization That Chronic Inflammation Is Crucial in Many Diseases Opens New Avenues for Treatment, *EMBO Reports*, 2012, 13(11):968-970.
161. www.sciencedirect.com/science/article/pii/S001650851201493X.
162. www.journals.lww.com/jwocnonline/Abstract/2002/07000/Irritable_Bowel_Syndrome__More_Than_a_Gut_Feeling.9.aspx.
163. www.jpp.krakow.pl/journal/archive/08_07_s3/pdf/131_08_07_s3_article.pdf.
164. Holzman D.C., What's in a Color? The Unique Human Health Effects of Blue Light, *Environmental Health Perspectives*, 2010, 118(1):A22-A27.

165. www.journals.plos.org/ploscompbiol/article?id=10.1371/journal.pcbi.0040004.
166. www.versatables.com/discover/sit-and-stand-movement/sedentary-lifestyles/sedentary-lifestyles-linked-to-depression-anxiety/.
167. www.ncbi.nlm.nih.gov/pubmed/11051338.
168. www.en.wikipedia.org/wiki/Acetylcholine.
169. www.jissn.biomedcentral.com/articles/10.1186/1550-2783-5-S1-P15.
170. www.ncbi.nlm.nih.gov/pmc/articles/PMC4527046/.
171. www.en.wikipedia.org/wiki/Cytokine.
172. www.ncbi.nlm.nih.gov/pubmed/18926158.
173. www.en.wikipedia.org/wiki/Gamma-Aminobutyric_acid.
174. Nuss P., Anxiety Disorders and GABA Neurotransmission: A Disturbance of Modulation, *Neuropsychiatric Disease and Treatment*, 2015, 11:165-175.
175. www.telegraph.co.uk/news/science/science-news/8316534/Welcome-to-the-information-age-174-newspapers-a-day.html.
176. www.cdc.gov/features/costsofdrinking/index.html.
177. www.ncbi.nlm.nih.gov/pubmed/19462324.
178. www.sciencedirect.com/science/article/pii/S0306987712000321.
179. www.en.wikipedia.org/wiki/Oxytocin.
180. www.ncbi.nlm.nih.gov/pmc/articles/PMC3537144/.
181. www.journal.frontiersin.org/article/10.3389/

fnhum.2015.00518/full.
182. www.sciencedirect.com/science/article/pii/014976349400070H.
183. Szeto A. et al., Oxytocin Attenuates NADPH-Dependent Superoxide Activity and IL-6 Secretion in Macrophages and Vascular Cells, *Physiol Endocrinol Metab*, 2008; 295(6):E1495-501.
184. www.en.wikipedia.org/wiki/Anandamide.
185. Kotler W., *The Rise of Superman*, 2014.
186. www.en.wikipedia.org/wiki/Arachidonic_acid.
187. www.bodybuilding.com/fun/arachidonic-acid-when-inflammation-is-good.html.
188. Mallet P.E. and Beninger R.J., The Endogenous Cannabinoid Receptor Agonist Anandamide Impairs Memory in Rats, *Behavioural Pharmacology*, 1996, 7 (3): 276-284.
189. Fuss J. et al., A runner's high depends on cannabinoid receptors in mice, *PNAS*, 2015, 112 (42): 13105-13108.
190. di Tomaso E., Beltramo M. and Piomelli D., Brain Cannabinoids in Chocolate, *Nature*, 1996 382 (6593): 677-8.
191. Ranabir S. and Reetu K., Stress and Hormones, *Indian Journal of Endocrinology and Metabolism*, 2011, 15(1):18-22.
192. www.ncbi.nlm.nih.gov/pubmed/22738346/#cm22738346_15964.
193. Tosini G., Ferguson I. and Tsubota K., Effects of Blue Light on the Circadian System and Eye Physiology, *Molecular Vision*,

2016, 22:61-72.
194. Masters A. et al., Melatonin, the Hormone of Darkness: From Sleep Promotion to Ebola Treatment, *Brain Disorders and Therapy*, 2014, 4(1):1000151.
195. www.scientificamerican.com/article/q-a-why-is-blue-light-before-bedtime-bad-for-sleep/.
196. www.nature.com/articles/srep11325.
197. Mead M.N., Benefits of Sunlight: A Bright Spot for Human Health, *Environmental Health Perspectives*, 2008, 116(4):A160-A167.
198. www.denverpost.com/2016/06/29/media-use-america-11-hours/.
199. My interview with Daniel Georgiev: www.ryanmunsey.com/132-beyond-blue-light-why-technology-is-killing-our-eyes-with-iris-creator-daniel-georgiev/.
200. www.jackkruse.com/reality-12-dopamine-blinds-us-from-natures-fractal-fabric/.
201. *Neuroscience Letters*, Elsevier.
202. *International Journal of Oncology*, Spandidos Publications.
203. Singer K., *An Electronic Silent Spring: Facing the Dangers and Creating Safe Limits*, 2014.
204. www.pnas.org/content/112/28/8567.abstract.
205. Kuo F.E. and Faber Taylor A., A Potential Natural Treatment for Attention-Deficit/Hyperactivity Disorder: Evidence from a National Study, *American Journal of Public Health*, 2004, 94(9):1580-1586.

206. www.apa.org/pubs/journals/releases/xlm-a0036577.pdf.
207. www.ncbi.nlm.nih.gov/pubmed/19121124.
208. www.pnas.org/content/112/28/8567.abstract.
209. www.ncbi.nlm.nih.gov/pubmed/16416750.
210. www.journals.plos.org/plosone/article?id=10.1371/journal.pone.0172200.
211. www.ncbi.nlm.nih.gov/pubmed/18332184.
212. www.sciencedirect.com/science/article/pii/S1875686709003327.
213. www.tandfonline.com/doi/abs/10.1080/036107399244093.
214. Buettner D., *The Blue Zones*, 2012.
215. www.ncbi.nlm.nih.gov/pubmed/22857379.
216. Chevalier G. et al., Health Implications of Reconnecting the Human Body to the Earth's Surface Electrons, *Journal of Environmental and Public Health*, 2012, 2012:291541.
217. www.earthinginstitute.net/grounding-improves-facial-abdominal-circulation/.
218. My interview with Jeff Nichols and Alexander Oliver: www.naturalstacks.com/blogs/news/77817157-how-navy-seals-are-using-float-tanks-to-treat-concussions.
219. Kotler W., *Stealing Fire*, 2017.
220. Kjellgren A. and Westman J., Beneficial Effects of Treatment with Sensory Isolation In Flotation-Tank As A Preventive Health-Care Intervention – A Randomized Controlled Pilot Trial, *BMC Complementary*

and Alternative Medicine, 2014, 14:417.
221. www.naturalstacks.com/blogs/news/112637381-floating-theta-states-recovery-with-sean-mccormick.
222. www.en.wikipedia.org/wiki/Alpha_wave#History_of_alpha_waves.
223. www.en.wikipedia.org/wiki/Hans_Berger.
224. www.en.wikipedia.org/wiki/Winfried_Otto_Schumann.
225. My interview with Dr. Andrew Hill: www.naturalstacks.com/blogs/news/get-peak-brain-performance-with-neuroscientist-dr-andrew-hill.
226. www.amadeux.net/sublimen/documenti/G.OsterAuditoryBeatsintheBrain.pdf.
227. www.sciencedaily.com/releases/2017/04/170412181341.htm.
228. www.academia.edu/258576/Crossmodal_transfer_of_emotion_by_music.
229. www.scientificamerican.com/article/smile-it-could-make-you-happier/.
230. www.masaru-emoto.net/english/water-crystal.html.
231. Hawkins D.R., *Power Vs Force*, 1994.
232. Roque A.L. et al., The Effects of Auditory Stimulation with Music on Heart Rate Variability in Healthy Women, *Clinics*, 2013, 68(7):960-967.
233. www.ncbi.nlm.nih.gov/pubmed/16038775.
234. www.news.wisc.edu/meditation-may-fine-tune-control-over-attention/.
235. Tyagi A. and Cohen M., Yoga and Heart

Rate Variability: A Comprehensive Review of The Literature, *International Journal of Yoga*, 2016, 9(2):97-113.
236. Khattab K. et al., Yoga Increases Cardiac Parasympathetic Nervous Modulation Among Healthy Yoga Practitioners, *Evidence-based Complementary and Alternative Medicine: eCAM*, 2007, 4(4):511-517.
237. www.ncbi.nlm.nih.gov/pubmed/15750381.
238. www.bu.edu/news/2012/03/07/researchers-find-yoga-helps-ease-stress-related-medical-and-psychological-conditions/.
239. www.ncbi.nlm.nih.gov/pubmed/18991518.
240. My interview with Brain.fm: www.naturalstacks.com/blogs/news/76646277-boost-focus-immediately-with-music-backed-by-neuroscience.
241. Shanahan C., *Deep Nutrition: Why Your Genes Need Traditional Food*, 2017.
242. Divine M., *The Way of the SEAL: Think Like an Elite Warrior to Lead and Succeed*, 2013.
243. Hof W., *Becoming the Iceman*, 2011.
244. www.en.wikipedia.org/wiki/Tummo.
245. www.ncbi.nlm.nih.gov/pubmed/11447037.
246. www.ncbi.nlm.nih.gov/pubmed/18785356.
247. Van der Lans A.A.J.J. et al., Cold Acclimation Recruits Human Brown Fat and Increases Nonshivering Thermogenesis, *The Journal of Clinical Investigation*, 2013, 123(8):3395-3403.
248. Kirsi A. et al., Functional Brown Adipose Tissue in Healthy Adults, *New England J Med*,

2009; 360:1518-1525.
249. www.ncbi.nlm.nih.gov/pubmed/6722594.
250. www.jappl.org/content/95/4/1584.full.
251. www.foundmyfitness.com/reports/cold-stress.pdf.
252. www.jackkruse.com/cold-thermogenesis-6-the-ancient-pathway/.
253. www.reportlinker.com/insight/smartphone-connection.html.
254. www.melrobbins.com/the-5-second-rule/
255. www.news.stanford.edu/2009/08/24/multitask-research-study-082409/.
256. www.reliableplant.com/Read/8259/fail-achieve-goals.
257. Covey S., *The 7 Habits of Highly Effective People*, 2004.
258. www.phys.org/news/2017-04-spent-online-children-happy.html.
259. www.guilfordjournals.com/doi/abs/10.1521/jscp.2014.33.8.701.
260. www.forbes.com/sites/amitchowdhry/2016/04/30/study-links-heavy-facebook-and-social-media-usage-to-depression/#5db8d0124b53.
261. www.mediakix.com/2016/12/how-much-time-is-spent-on-social-media-lifetime/#gs.8=dblpY.
262. www.opinionator.blogs.nytimes.com/2012/01/22/anatomy-of-fear/?_r=0
263. Steimer T., The Biology of Fear and Anxiety-Related Behavior, *Dialogues in Clinical Neuroscience*, 2002, 4(3):231-249.

264. Ropeik D., The consequences of Fear, *EMBO Reports*, 2004, 5(Suppl 1):S56-S60.
265. Johnson L., How Fear and Stress Shape the Mind, *Frontiers in Behavioral Neuroscience*, 2016.
266. Radley J. J., Toward A Limbic Cortical Inhibitory Network: Implications for Hypothalamic-Pituitary-Adrenal Responses Following Chronic Stress, *Frontiers in Behavioral Neuroscience*, 29 March 2012.
267. Dr. Mckay's TED talk: www.youtube.com/watch?v=xiXZVDKRe00.
268. Wenner M., Smile! It Could Make You Happier, *Scientific American*, 9/1/2009.
269. Lerner J., Emotion and Decision Making, Harvard University, *Annual Review of Psychology*, 16 June 2014
270. Oppong T. Accountability Accelerates Your Performance. https://medium.com/the-mission/the-accountability-effect-a-simple-way-to-achieve-your-goals-and-boost-your-performance-8a07c76ef53a.

About The Author

The co-founder and podcast host at the Better Human Project, Ryan is a high performance consultant, author, and speaker. A scientist by training, he has also worked as a fitness model, nutritionist, gym owner, and corporate wellness consultant.

He currently lives in his home state of Virginia with his wife Donna who is an Internal Medicine Physician.

Ryan's consulting clients include Olympic athletes, special forces operators, entrepreneurs and C-level executives around the world. For more information, on hiring Ryan as a coach, consultant or speaker for your next event, visit ryanmunsey.com. You can also find dates & locations for Ryan's workshops at ryanmunsey.com

Made in United States
North Haven, CT
02 November 2022